George Payn Quackenbos

A Mental Arithmetic

George Payn Quackenbos

A Mental Arithmetic

ISBN/EAN: 9783743393646

Manufactured in Europe, USA, Canada, Australia, Japa

Cover: Foto ©berggeist007 / pixelio.de

Manufactured and distributed by brebook publishing software (www.brebook.com)

George Payn Quackenbos

A Mental Arithmetic

APPLETONS' MATHEMATICAL SERIES.

A
MENTAL ARITHMETIC.

BY

G. P. QUACKENBOS, LL. D.,

AUTHOR OF

"AN ENGLISH GRAMMAR;" "FIRST LESSONS IN COMPOSITION;" "ADVANCED COURSE OF COMPOSITION AND RHETORIC;" "A NATURAL PHILOSOPHY;" "ILLUSTRATED SCHOOL HISTORY OF THE UNITED STATES;" "PRIMARY HISTORY OF THE UNITED STATES," ETC.

NEW YORK:
D. APPLETON AND COMPANY.
549 & 551 BROADWAY.
1878.

[ADVERTISEMENT.]

APPLETONS' MATHEMATICAL SERIES.

BY G. P. QUACKENBOS, LL. D.,

A. Primary Arithmetic. Beautifully illustrated; carries the beginner through the first four Rules and the simple Tables, combining mental exercises with examples for the slate. 16mo. 108 pages. 22 cents.

An Elementary Arithmetic. Reviews the subjects of the Primary in a style adapted to somewhat maturer minds. Also embraces Fractions, Federal Money, Reduction, and the Compound Rules. 12mo. 144 pages. 40 cents.

A Practical Arithmetic. Prepared with direct reference to the wants of Common Schools, giving special prominence to the branches of Mercantile Arithmetic. 12mo. 336 pages. 80 cents

A Mental Arithmetic. Designed to impart readiness in mental calculations, and extending them to the various operations needed in business life. Introduces short methods, and new and beautiful processes. 16mo. 168 pages. 35 cents.

A Higher Arithmetic. 12mo. 420 pages. $1.10.

Entered, according to Act of Congress, in the year 1868, by
D. APPLETON & CO.,
In the Clerk's Office of the District Court of the United States for the
Southern District of New York.

PREFACE.

THE study of Mental Arithmetic has two principal objects in view, to discipline the mind and impart quickness and accuracy in mental calculations. To secure these objects in the highest degree and in the pleasantest way to both teacher and learner is the aim of this little volume. It is intended for pupils who have gone through a Primary Arithmetic, and know how to read and write numbers up to thousands inclusive; and may be used either by itself to succeed the Primary or Elementary, or as an auxiliary to the Elementary or Practical, on alternate days.

Among the more important features of the present work are the following:—1. The gradual and inductive mode of unfolding the subject, with the aid of rigid but clear analyses expressed as briefly as possible. 2. The introduction of necessary definitions, too often excluded from Mental Arithmetics. 3. The great variety and practical bearing of the Examples; the careful avoidance of obscurity in wording them, and the exclusion of all questions involving impossibilities or absurdities. 4. The presentation of the Metric System, hitherto confined mainly to text-books on written Arithmetic. 5. The teaching of short methods and processes actually used in the counting-room. 6. The extension of mental calculations to more of the operations of every-day business life than has hitherto been attempted; such as equation of payments, stock jobbing, U. S. securities, taxes, duties, &c. The value of this last feature, it is believed, can hardly fail to be appreciated in this practical age.

The interest and profit with which classes will use this work will depend entirely on the thoroughness with which the successive steps are taken. Review again and again if necessary, and let nothing pass till it is mastered. Short lessons should be given, to be prepared beforehand. The books should be closed at recitation, the question read but once, and the scholars should have no intimation as to which of their number will be called on to solve it. The analyses given in the Models should be followed (unless better ones can be devised), with distinct articulation and in correct language. Let the answer always be distinctly stated, when it is reached, in connection with the denomination, as in the Model. Whenever any particular form of analysis has become perfectly familiar, it will be well to omit it in the case of some of the questions, and require immediate answers, as well to encourage quickness of thought as to economize time. A few questions from previous lessons, to be answered thus promptly, will be found useful at each recitation.

The Author can only hope that this work may meet with as cordial a reception as has been so kindly extended to the other Numbers of the Series.

NEW YORK, *May* 22, 1868.

CONTENTS.

		PAGE
CHAPTER FIRST,	ADDITION,	5
CHAPTER SECOND,	SUBTRACTION,	13
CHAPTER THIRD,	MULTIPLICATION,	20
CHAPTER FOURTH,	DIVISION,	28
CHAPTER FIFTH,	FRACTIONS,	38
CHAPTER SIXTH,	FEDERAL MONEY,	72
CHAPTER SEVENTH,	REDUCTION,	75
CHAPTER EIGHTH,	THE METRIC SYSTEM,	89
CHAPTER NINTH,	THE COMPOUND RULES,	94
CHAPTER TENTH,	MISCELLANEOUS EXAMPLES,	100
CHAPTER ELEVENTH,	PERCENTAGE,	117
CHAPTER TWELFTH,	INTEREST,	127
CHAPTER THIRTEENTH,	DISCOUNT,	140
CHAPTER FOURTEENTH,	STOCKS,—U. S. SECURITIES,	143
CHAPTER FIFTEENTH,	MISCELLANEOUS EXAMPLES,	150

MENTAL ARITHMETIC.

CHAPTER FIRST.

ADDITION.

[It is supposed that the pupil has learned the Addition, Subtraction, Multiplication, and Division Tables. Let them be reviewed, as presented under Chapters 1, 2, 3, and 4, until he can say them promptly and perfectly, backward as well as forward.]

Addition Table.

1 and	2 and	3 and	4 and	5 and
1 are 2	1 are 3	1 are 4	1 are 5	1 are 6
2 are 3	2 are 4	2 are 5	2 are 6	2 are 7
3 are 4	3 are 5	3 are 6	3 are 7	3 are 8
4 are 5	4 are 6	4 are 7	4 are 8	4 are 9
5 are 6	5 are 7	5 are 8	5 are 9	5 are 10
6 are 7	6 are 8	6 are 9	6 are 10	6 are 11
7 are 8	7 are 9	7 are 10	7 are 11	7 are 12
8 are 9	8 are 10	8 are 11	8 are 12	8 are 13
9 are 10	9 are 11	9 are 12	9 are 13	9 are 14
10 are 11	10 are 12	10 are 13	10 are 14	10 are 15

6 and	7 and	8 and	9 and	10 and
1 are 7	1 are 8	1 are 9	1 are 10	1 are 11
2 are 8	2 are 9	2 are 10	2 are 11	2 are 12
3 are 9	3 are 10	3 are 11	3 are 12	3 are 13
4 are 10	4 are 11	4 are 12	4 are 13	4 are 14
5 are 11	5 are 12	5 are 13	5 are 14	5 are 15
6 are 12	6 are 13	6 are 14	6 are 15	6 are 16
7 are 13	7 are 14	7 are 15	7 are 16	7 are 17
8 are 14	8 are 15	8 are 16	8 are 17	8 are 18
9 are 15	9 are 16	9 are 17	9 are 18	9 are 19
10 are 16	10 are 17	10 are 18	10 are 19	10 are 20

ADDITION.

SECTION 1.—**Addition** is the process of uniting two or more numbers in one, called their **Sum**. *Two* and *one* are *three;* we have *added* 2 and 1, and 3 is their *sum*.

Addition is denoted by an erect cross +, called **Plus**, placed between the numbers to be added. $2+1$ is read *two plus one*, and means that 2 and 1 are to be added.

Two short horizontal lines = denote equality. $2+1=3$, is read *two plus one equals three*, and means that the sum of 2 and 1 is 3.

1. What is the sum of 2 and 3? 2 and 9?
2. What is the sum of 2 and 4? 2 and 7?
3. What is the sum of 3 and 1? 3 and 4?
4. What is the sum of 3 and 8? 3 and 6?
5. What is the sum of 4 and 5? 4 and 2?
6. What is the sum of 4 and 9? 4 and 10?
7. How many are 5 and 8? 5 and 1? 5 and 5?
8. How many are 5 and 9? 5 and 3? 5 and 6?
9. How many are 6 and 2? 6 and 9? 6 and 5?
10. How many are 6 and 1? 6 and 6? 6 and 8?
11. Add 7 and 7. 7 and 2. 7 and 4. 7 and 9.
12. Add 7 and 8. 7 and 1. 7 and 5. 7 and 10.
13. Add 8 and 8. 8 and 6. 8 and 3. 8 and 4.
14. Add 8 and 1. 8 and 9. 8 and 2. 8 and 7.
15. $9+4=$ how many? $9+9$? $9+1$? $9+0$?
16. $9+7=$ how many? $9+2$? $9+6$? $9+3$?
17. What is the value of $10+1$? $10+10$? $10+9$? $10+5$? $5+10$? $10+8$? $10+4$? $1+9$? $1+5$?
18. How many are 2 and 5? 5 and 2? How much is $6+4$? $4+6$? $3+7$? $7+3$?

ADDITION.

19. When two numbers are to be added, does it make any difference which is taken first?

20. How many are 4 and 8? How many are 1 and 3 and 8? How many are 2 and 2 and 8?

21. How many are 3 and 9? How many are 1 and 2 and 9? How many are 2 and 1 and 9?

22. How many are 8 and 10? How much is $3+5+10$? How much is $2+2+4+10$?

23. How many are 6 and 7? How much is $3+3+7$? How much is $1+4+1+7$?

24. How many are 9 and 8? How much is $5+4+8$? How much is $3+5+1+8$?

25. How many are 8 and 5? How many are 8 and $3+2$? How many are 8 and $4+1$?

26. How many are 10 and 7? $2+8$ and $4+3$? 10 and $1+6$? $1+6+10$?

27. Mary and her seven sisters just filled a bench; how many did the bench hold?

28. You have four fingers on each hand; how many fingers have you on both hands?

29. A newsboy bought five daily and seven weekly papers; how many papers had he in all?

30. If Ruth has seven pins in one pin-cushion, and six in another, how many pins has she in both?

31. Guy pulled six ears of corn on Monday, and ten on Tuesday; how many did he pull both days?

32. Ten birds were sitting on a steeple, when six more alighted; how many were there then?

33. Henry had two cents, and his father gave him ten more. Louis had ten cents, and he found two more in the street. Which then had the most?

34. Put four marbles in a bag already containing ten, and how many will you have in the bag?

35. A boy caught six perch, three pickerel, and ten shiners; how many fish did he catch in all?

36. A baker gave nine loaves to one poor family, and five to another; how many did he give both?

37. How many trees stand beside my lane, if there are ten on one side and three on the other?

38. A mother, having three sons aged two, five, and eight years, gave each as many dollars as he was years old; how many dollars did she give all three?

39. How many books had Paul, if his father gave him 3, his mother 6, his brother 1, and his sister 7?

40. A farmer had four ducks, six geese, and as many chickens as he had ducks and geese put together; how many fowls had he in all?

SECTION 2.—1. How many are 3 and 2? 13 and 2? 33 and 2? 63 and 2? 93 and 2? 103 and 2?

2. What does $2+3$ equal? $22+3$? $43+2$? $53+2$? $83+2$? $183+2$? $193+2$? $203+2$?

3. What is the sum of 4 and 5? 5 and 4? 55 and 4? 54 and 5? 74 and 5? 75 and 4?

4. How much is $7+2$? $107+2$? $207+2$? $507+2$? $807+2$? $1007+2$? $4007+2$? $5007+2$?

5. Add 2 and 6. 62 and 6. 162 and 6. 1062 and 6. 1066 and 2. 166 and 2. 66 and 2. 76 and 2.

6. How much is $3+4$? $20+3+4$? $30+3+4$? $40+3+4$? $3+40+4$? $60+4+3$? $4+3+70$?

7. How much is $5+2$? $50+5+2$? $51+5+2$?

ADDITION.

8. How many are 9 and 1? 89 and 1? 99 and 1? 109 and 1? 199 and 1? 209 and 1? 299 and 1?

9. How many are 6 and 4? 6 and 5? 16 and 4? 16 and 5? 26 and 5? 36 and 5? 56 and 5?

10. How many are 8 and 7? 18 and 7? 118 and 7? 128 and 7? 28 and 7? 38 and 7? 638 and 7?

11. What is $3+9$ equal to? $4+9$? $23+9$? $24+9$? $54+9$? $55+9$? $56+9$? $52+9$?

12. What is $2+5+7$ equal to? $12+5+7$? $22+5+7$? $42+5+7$? $72+5+7$? $92+5+7$?

13. How many are 63 and 5? 79 and 2? 103 and 8? 47 and 6? 99 and 3? 102 and 7? 113 and 5?

14. How many are 10 and 10? 20 and 10? 50 and 10? 90 and 10? 100 and 10? 200 and 10?

15. 6 and 9 make how many? 34 and 6? 43 and 9? 52 and 8? 58 and 2? 50 and 3? 79 and 6? 88 and 4? 103 and 3? 111 and 9? 124 and 4?

16. How many are 7 and 7 and 8? 21 and 9 and 6? 34 and 8 and 7? 41 and 10 and 9? 72 and 4 and 5? 99 and 10 and 1? 199 and 2 and 9?

17. One tree bears 34 apples, another 6, and a third 10; how many apples do all three bear?

MODEL. If one tree bears 34 apples, another 6, and a third 10, all three together will bear the sum of 34, 6, and 10 apples, or 50 apples. *Answer*, 50 apples.

18. A farmer set out 42 apple, 8 pear, and 9 plum trees; how many trees did he set out altogether?

19. If I travel 75 miles by boat, 7 by railroad, and 10 by stage, how many miles do I go in all?

20. 103 English books, 20 French books, and 6 German books, make how many books in all?

21. James bought 20 cents' worth of bread, 10 cents' worth of cake, and 9 cents' worth of crackers; how much did he have to pay the baker?

22. How many miles will you travel in a day, if you go 21 miles in the morning, 30 miles in the afternoon, and 8 miles in the evening?

23. In a field were 19 sheep, 8 cows, and 5 calves; how many animals were in the field?

24. John was eight years older than Charles. How old was John when Charles was thirteen? How old was each, nine years afterward?

25. How many windows in a factory, if there are 12 in front and 6 on each of the other three sides?

26. In a certain school were 10 boys and 6 more girls than boys; how many scholars altogether?

27. Helen laid out 45 cents for muslin, 5 cents for thread, 9 cents for pins, and 6 cents for needles; how many cents did she spend in all?

28. A fisherman had sold all his stock, except 17 shad, 6 bass, and 9 eels; how many fish had he left?

29. Count by 2's, beginning 2, 4, 6, &c., up to 100.
30. Count by 2's, beginning 1, 3, 5, &c., up to 99.
31. Count by 3's, beginning 1, 4, 7, &c., up to 100.
32. Count by 4's, beginning 2, 6, 10, &c., up to 98.
33. Count by 5's, beginning 1, 6, 11, &c., up to 96.
34. Count by 6's, beginning 2, 8, 14, &c., up to 98.
35. Count by 7's, beginning 3, 10, 17, &c., up to 94.
36. Count by 8's, beginning 5, 13, 21, &c., up to 93.
37. Count by 9's, beginning 1, 10, 19, &c., up to 100.

[These exercises may be continued, and varied by commencing differently, till they are made perfectly familiar.]

ADDITION.

SECTION 3.—1. How many are 43 and 26?

MODEL. 43=3 units 4 tens; 26=6 units 2 tens. 6 units and 3 units are 9 units; 2 tens and 4 tens are 6 tens. 6 tens 9 units are 69. *Ans.* 69.

2. What is the sum of 31 and 52? Of 74 and 15?
3. How many are 22 and 60? 17+81? 65+32?
4. How many are 14 and 54? 46+31? 80+19?
5. Add 123 and 876. 456 and 543. 739 and 40.
6. What is the sum of 267 and 431?

SHORT FORM. 1 and 7 are 8; 3 and 6 are 9; 4 and 2 are 6. *Ans.* 698.

7. How many are 240 and 326? 579 and 120?
8. How many are 165 and 722? 410 and 378?
9. How many are 821 and 73? 26 and 933?
10. What is the sum of 59 and 63?

SHORT FORM. 3 and 9 are 12, 2 units and 1 ten; 1 and 6 and 5 are 12. *Ans.* 122.

11. How many are 14 and 18? 17+23? 21+16?
12. How many are 40 and 69? 53+29? 37+47?
13. How many are 99 and 12? 86+62? 29+59?
14. How many are 108 and 15? 221+49? 72+68?
15. How many are 29 and 29? 38+57? 93+67?
16. Count by tens,—10, 20, 30, &c.,—up to 100.
17. Count by elevens,—11, 22, 33, &c.,—up to 132.
18. Count by twelves,—12, 24, 36, &c.,—up to 144.
19. Andrew had 23 marbles, and won 38 more; how many had he then?
20. If I give 57 cents to one poor family, and 39 cents to another, how much do I give both?
21. Which is greater, 26+57 or 77+16?
22. After spending 18 dollars for groceries, and 27

dollars for drygoods, a lady found that she had five dollars left. How many dollars had she at first?

23. How many birds are in two flocks, one containing 97 birds, and the other 64?

24. A drover who had 56 sheep, bought 14 more, and another flock of 30; how many sheep had he then?

25. Noah lived 600 years before the Flood, and 350 years after it; how old was Noah when he died?

26. Methuselah lived 19 years more than Noah; what age did Methuselah attain?

27. A man who gave 125 dollars for a watch, sold it for 25 dollars more than it cost; what did he get for it?

28. How many pounds are there in two bales of cotton, one weighing 404 pounds, and the other 382?

29. If a farmer raises 93 bushels of wheat and 128 of corn, how many bushels of both does he raise?

30. A lot was bought for 360 dollars, and sold at a profit of 75 dollars; how much did it bring?

MODEL. If it was bought for 360 dollars and sold at a profit of 75 dollars, it must have brought the sum of 360 and 75 dollars, or 435 dollars. *Ans.* 435 dollars.

31. A horse was bought for 238 dollars, and sold at a profit of 27 dollars; what was it sold for?

32. A man makes 56 dollars by selling some goods that cost him 249 dollars; what does he sell them for?

33. An orchard cost 275 dollars; the profit on it being 87 dollars, what was it sold for?

34. Conrad had 8 books, containing 63 pictures, and 19 books, containing 148 pictures. How many books had Conrad in all, and how many pictures?

CHAPTER SECOND.

SUBTRACTION.

SECTION 4.—**Subtraction** is the process of taking one number from another.

Subtraction is denoted by a short horizontal line —, called **Minus**, placed before the smaller number.

4—1 is read *four minus one*, and means that 1 is to be subtracted from 4.

4—1=3. We have subtracted 1 from 4; the result, 3, is called the **Remainder** or **Difference**.

SUBTRACTION TABLE.

1 from	2 from	3 from	4 from	5 from
1 leaves 0	2 leaves 0	3 leaves 0	4 leaves 0	5 leaves 0
2 leaves 1	3 leaves 1	4 leaves 1	5 leaves 1	6 leaves 1
3 leaves 2	4 leaves 2	5 leaves 2	6 leaves 2	7 leaves 2
4 leaves 3	5 leaves 3	6 leaves 3	7 leaves 3	8 leaves 3
5 leaves 4	6 leaves 4	7 leaves 4	8 leaves 4	9 leaves 4
6 leaves 5	7 leaves 5	8 leaves 5	9 leaves 5	10 leaves 5
7 leaves 6	8 leaves 6	9 leaves 6	10 leaves 6	11 leaves 6
8 leaves 7	9 leaves 7	10 leaves 7	11 leaves 7	12 leaves 7
9 leaves 8	10 leaves 8	11 leaves 8	12 leaves 8	13 leaves 8
10 leaves 9	11 leaves 9	12 leaves 9	13 leaves 9	14 leaves 9
6 from	**7 from**	**8 from**	**9 from**	**10 from**
6 leaves 0	7 leaves 0	8 leaves 0	9 leaves 0	10 leaves 0
7 leaves 1	8 leaves 1	9 leaves 1	10 leaves 1	11 leaves 1
8 leaves 2	9 leaves 2	10 leaves 2	11 leaves 2	12 leaves 2
9 leaves 3	10 leaves 3	11 leaves 3	12 leaves 3	13 leaves 3
10 leaves 4	11 leaves 4	12 leaves 4	13 leaves 4	14 leaves 4
11 leaves 5	12 leaves 5	13 leaves 5	14 leaves 5	15 leaves 5
12 leaves 6	13 leaves 6	14 leaves 6	15 leaves 6	16 leaves 6
13 leaves 7	14 leaves 7	15 leaves 7	16 leaves 7	17 leaves 7
14 leaves 8	15 leaves 8	16 leaves 8	17 leaves 8	18 leaves 8
15 leaves 9	16 leaves 9	17 leaves 9	18 leaves 9	19 leaves 9

SUBTRACTION.

1. How many does 3 from 5 leave? 4 from 8? 7 from 9? 1 from 4? 2 from 6? 5 from 7?

2. Take 1 from 9. 2 from 10. 3 from 11. 9 from 18. 7 from 14. 10 from 15. 6 from 15. 4 from 13.

3. How much is 7 less 4? 17 less 4? 27 less 4? 47 less 4? 87 less 4? 37 less 4? 97 less 4?

4. How much is 8—2? 98—2? 108—2? 68—2? 168—2? 268—2? 768—2? 778—2?

5. Subtract 7 from 9. 27 from 29. 97 from 99. 107 from 109. 7 from 8. 207 from 208.

6. How much is 11—10? 31—10? 51—10? 81—10? 13—10? 23—10? 123—10? 223—10?

7. How much remains, if we take 20 from 40? 20 from 47? 20 from 67? 20 from 89? 20 from 39?

8. 30 from 50 leaves how many? 50 from 90? 50 from 91? 60 from 82? 40 from 78? 70 from 83?

9. 21 from 43? 56 from 67? 71 from 85? 38 from 59? 86 from 98? 31 from 77? 44 from 59?

10. If a man bought a lamp for 5 dollars and sold it for 8, how many dollars did he make?

MODEL. If a man bought a lamp for 5 dollars and sold it for 8, he must have made the difference between 5 and 8 dollars, or 3 dollars. *Ans.* 3 dollars.

11. What is the profit on a barrel of flour, bought for 8 dollars and sold for 12 dollars?

12. Mary is 13 years old and Sarah 19; two years hence, what will be the difference in their ages?

13. How many more boys than girls are there in a school, if there are 75 boys and 32 girls?

14. In a stable there were 23 horses and 46 stalls; how many empty stalls were there?

15. A gentleman now 87 years old was married 56 years ago; how old was he when he was married?

16. A boy carrying 39 eggs in a basket, fell and broke 19 of them; how many were left whole?

17. Ninety-three persons were wrecked in a storm. Eighty were saved; how many perished?

18. A person who had 88 melons, sold 67 of them; how many had he left?

19. How much does a merchant make on goods bought for 68 dollars and sold for 88?

20. A drover, having 96 sheep, divided them into 2 flocks. He put 25 in one flock; how many in the other? If from the second flock he sold 40, how many had he left in that flock? How many in all?

21. What change must a boy, who pays 15 cents for a ball, get for a 25-cent stamp?

SECTION 5.—1. How much does 3 from 12 leave? 3 from 22? 3 from 32? 3 from 52? 3 from 72?

2. 5 from 11 leaves how many? 5 from 21? 5 from 51? 5 from 91? 5 from 31? 5 from 71?

3. 7 from 10? 7 from 20? 7 from 80? 7 from 60? 7 from 30? 7 from 70? 7 from 50? 7 from 40?

4. 8 from 14? 8 from 54? 8 from 94? 8 from 24? 2 from 11? 2 from 71? 2 from 31? 2 from 81?

5. How much is 10—9? 30—9? 60—9? 13—4? 93—4? 83—4? 83—5? 13—5?

6. How much is 12—6? 42—6? 62—6? 17—9? 57—9? 37—9? 10—2? 90—2? 100—2? 200—2? 16—8? 100—8? 300—8? 12—7? 92—7? 102—7?

7. Take 5 from 10. 5 from 30. 5 from 130. 5 from 230. 6 from 14. 6 from 114. 6 from 214.

8. How much is 11 less 7? 111 less 7? 411 less 7? 15 less 9? 115 less 9? 215 less 9? 25 less 9?

9. How much does 8 from 12 leave? 8 from 42? 8 from 142? 3 from 11? 3 from 81? 3 from 101?

10. 34 from 41 leaves how many? 24 from 51? 64 from 91? 14 from 71? 54 from 81?

11. 15 from 32? 35 from 72? 55 from 92? 26 from 33? 46 from 63? 16 from 93? 86 from 103?

12. 37 from 94? 67 from 84? 18 from 45? 28 from 85? 4 from 10? 4 from 100? 24 from 70?

13. How much is 17−8? 47−18? 67−28? 87−38? 42−5? 42−25? 72−55? 92−35?

14. Take 14 from 42. 114 from 142. 34 from 142. 44 from 152. 25 from 34. 65 from 84.

15. 9 from 13 leaves how many? 29 from 53? 11 from 60? 13 from 80? 41 from 90? 74 from 80? 26 from 90? 17 from 36? 49 from 93? 29 from 71?

16. When subtraction is to be denoted, what sign is used? Which number is it placed before?

17. What sign is used, when addition is to be denoted? What is the sign of equality?

18. Count backward by 2's from 100: 100, 98, &c.
19. Count backward by 3's from 100: 100, 97, &c.
20. Count backward by 4's from 100: 100, 96, &c.
21. Count backward by 4's from 99: 99, 95, 91, &c.
22. Count backward by 5's from 100: 100, 95, &c.
23. Count backward by 6's from 100; from 95.
24. Count backward by 7's from 100; from 98.
25. Count backward by 8's from 100; by 9's.

SUBTRACTION.

26. A man bought a cow for $75,* and paid $29 on account. How much remained due?

MODEL. If a man bought a cow for $75 and paid $29 on account, there remained due the difference between $75 and $29, or $46. *Ans.* $46.

27. From New York to Troy by the Hudson River Railroad is 150 miles; from New York to Poughkeepsie by the same road is 74 miles. How far is it from Poughkeepsie to Troy?

28. A owes B $47, and pays him on account $18; how much does he then owe B?

29. If 63 melons were stolen from a pile containing 100, how many were left?

30. If you have used 88 pens out of a box containing 144, how many are left?

31. A pole 72 inches long is driven into the ground 24 inches; how many inches of the pole are left above ground? If 14 inches are then cut off, how many will be left above ground?

32. C owes D $78. If he hands D a hundred-dollar bill, how much change should he receive?

33. If 17 pounds of butter are used out of a firkin that held 63 pounds, how many pounds are left?

34. Charles is 25 years younger than his father; how old is he, when his father is 73?

35. A hardware merchant sold for $101 some iron that cost $88; what was his profit?

36. What number must I add to 111, to produce 132? To produce 144?

---

* This mark ($) denotes dollars. It is always placed before the number. $75 is read *seventy-five dollars.*

SECTION 6.—1. How much is $8+9-7$?

MODEL. $8+9$ equals 17. $17-7$ equals 10. *Ans.* 10.

2. How much is $6+13-8$? $14+8-12$?
3. How much is $15-9+8$? $23-11+7$?
4. How much is $34+12-9$? $47+11-12$?
5. How much is $23+45-32$? $56+12-35$?
6. How much is $31+17-25$? $73+26-52$?
7. How much is $42+39-28$? $67+24-19$?
8. How much is $76+22-49$? $18+56-36$?
9. From the sum of 47 and 58 subtract the sum of 57 and 48.
10. From the sum of 26 and 36 subtract the difference between 50 and 25.
11. From the difference between 26 and 88 take the sum of 17 and 27.
12. From the difference between 101 and 11 take the difference between 98 and 77.
13. From the difference between 84 and 16 take the difference between 45 and 36.
14. After a battle, 18 men in a certain company were found to be killed, and 34 wounded. If the company contained 89 men, how many were uninjured?

MODEL. If 18 men were killed, and 34 wounded, the total of killed and wounded was $18+34$, or 52, men; and the number uninjured, was the difference between 89 and 52, or 37, men. *Ans.* 37 men.

15. A person spent $37 for carpeting and $56 for furniture; what did the whole cost him? If he sold the whole for $100, what was his profit?
16. If a man who has $54 earns $21 more, and then spends $43, how much has he left?

SUBTRACTION.

17. James had 96 marbles; he gave away 24, and lost 36. How many had he left?

18. A certain boat has a crew of 18 persons, and 73 passengers on board. If 53 persons get off, how many are left aboard?

19. P had $20, and earned $13 more. Q had $46, and spent $19. Which then had the most, and how much?

20. An orchard of apple, pear, and cherry trees, contains 200 trees in all. If there are 125 apple trees, and 46 pear trees, how many cherry trees are there?

21. A farmer, who had made 336 pounds of butter, sold 110 pounds to one customer, and 127 to another; how many pounds were left?

22. If a person gets 14 quarts of strawberries from one bed, 19 from another, and 23 from a third, how many quarts will he have left after selling 37 quarts?

23. If a person gets 29 quarts of strawberries from one bed and 38 from another, how many quarts will he have left after selling 17 quarts and giving 25 away?

24. A and B start from two places 96 miles apart, and travel toward each other. The first day, A goes 48 miles, and B 27; how many miles are they then apart? How far is A from B's starting-place? How far is B from A's starting-place?

25. What number must you add to the sum of nine, thirteen, and forty-eight, to produce 99?

26. There are sixty-six rows of corn to hoe in one field, and forty-three in another. When Robert has hoed 19 rows, Richard 23, and Reuben 27, how many rows still remain to be hoed?

CHAPTER THIRD.

MULTIPLICATION.

SECTION 7.—Multiplication is the process of taking a number a certain number of times.

Multiplication is denoted by a slanting cross ×, placed between the numbers to be multiplied together.

MULTIPLICATION TABLE.

Once	Twice	3 times	4 times	5 times	6 times
1 is 1	1 is 2	1 is 3	1 is 4	1 is 5	1 is 6
2 is 2	2 is 4	2 is 6	2 is 8	2 is 10	2 is 12
3 is 3	3 is 6	3 is 9	3 is 12	3 is 15	3 is 18
4 is 4	4 is 8	4 is 12	4 is 16	4 is 20	4 is 24
5 is 5	5 is 10	5 is 15	5 is 20	5 is 25	5 is 30
6 is 6	6 is 12	6 is 18	6 is 24	6 is 30	6 is 36
7 is 7	7 is 14	7 is 21	7 is 28	7 is 35	7 is 42
8 is 8	8 is 16	8 is 24	8 is 32	8 is 40	8 is 48
9 is 9	9 is 18	9 is 27	9 is 36	9 is 45	9 is 54
10 is 10	10 is 20	10 is 30	10 is 40	10 is 50	10 is 60
11 is 11	11 is 22	11 is 33	11 is 44	11 is 55	11 is 66
12 is 12	12 is 24	12 is 36	12 is 48	12 is 60	12 is 72

7 times	8 times	9 times	10 times	11 times	12 times
1 is 7	1 is 8	1 is 9	1 is 10	1 is 11	1 is 12
2 is 14	2 is 16	2 is 18	2 is 20	2 is 22	2 is 24
3 is 21	3 is 24	3 is 27	3 is 30	3 is 33	3 is 36
4 is 28	4 is 32	4 is 36	4 is 40	4 is 44	4 is 48
5 is 35	5 is 40	5 is 45	5 is 50	5 is 55	5 is 60
6 is 42	6 is 48	6 is 54	6 is 60	6 is 66	6 is 72
7 is 49	7 is 56	7 is 63	7 is 70	7 is 77	7 is 84
8 is 56	8 is 64	8 is 72	8 is 80	8 is 88	8 is 96
9 is 63	9 is 72	9 is 81	9 is 90	9 is 99	9 is 108
10 is 70	10 is 80	10 is 90	10 is 100	10 is 110	10 is 120
11 is 77	11 is 88	11 is 99	11 is 110	11 is 121	11 is 132
12 is 84	12 is 96	12 is 108	12 is 120	12 is 132	12 is 144

MULTIPLICATION.   21

2 × 3 is read *two multiplied by three*, and means two taken 3 times.

2 × 3 = 6. We have multiplied 2 by 3; the result, 6, is called their **Product**.

1. How much is 3 times 4? 4 times 9? Twice 12? 4 times 6? 9 times 2? 8 times 4?

2. What is the product of 4 and 2? 9 and 6? 6 and 7? 11 and 3? 11 and 4? 12 and 6? 8 and 3?

3. How much is 12 × 5? 6 × 8? 3 × 9? 8 × 7? 4 × 10? 10 × 6? 5 × 11? 12 × 3? 7 × 7?

4. Multiply 8 by 8. 4 by 4. 6 by 6. 11 by 11. 5 by 5. 3 by 3. 12 by 12. 2 by 2.

5. How much is 12 times 7? 7 times 12? 6 times 11? 11 times 6? 3 × 4 × 2? 2 × 4 × 3?

6. When numbers are to be multiplied together, does the order in which they are taken affect the product?

7. How much is 7 times 9? 9 times 7? Twice 6? 6 times 2? 12 × 4? 4 × 12? 3 × 11? 11 × 3?

8. How much is 10 × 9? 6 × 10? 10 × 7? 2 × 10? 10 × 5? 8 × 10? 10 × 10? 10 × 3?

9. When 10 is a factor, with what figure does the product always end?

10. How much is 5 × 1? 5 × 9? 5 × 3? 5 × 7? 5 × 2? 5 × 6? 5 × 8? 6 × 5? 9 × 5?

11. When 5 is a factor, with one of what two figures does the product end?

12. What will 9 tops cost, at 5 cents apiece?

MODEL. Nine tops will cost 9 times as much as 1 top. If 1 top costs 5 cents, 9 tops will cost 9 times 5, or 45, cents. *Ans.* 45 cents.

13. What will 8 knives cost, at 6 dimes each?

14. At 12 cents each, what will 9 primers cost?

15. At 10 dollars each, what will 11 muffs cost?

16. What cost 8 oranges, at 3 cents apiece?

17. What cost 7 melons, at 9 cents apiece?

18. What cost 12 almanacs, at 6 cents each?

19. If a boat moves at the rate of 11 miles an hour, how far will it go in 11 hours?

MODEL. In eleven hours it will go 11 times as many miles as in 1 hour. If in 1 hour it goes 11 miles, in 11 hours it will go 11 times 11, or 121, miles. *Ans.* 121 miles.

20. How far will a stage, moving at the rate of 7 miles an hour, go in 6 hours? In 10 hours?

21. How many apples in 5 baskets, each containing 12 apples?

22. How many fingers have 11 persons?

23. How many thumbs have 8 persons?

24. If a boy earns $10 a week, how many dollars will he earn in 4 weeks? In 6 weeks? In 9 weeks?

25. How many chickens in eleven broods, containing nine chickens each?

26. Paid $5 for a lamp, and $1 for a shade. How much will four such lamps and shades cost?

27. At the rate of 7 dimes for a knife, and 2 dimes for a fork, what will 12 knives and forks cost?

28. Four pens, each containing 5 pigs, hold how many pigs in all?

29. $5 \times 4 =$ how many?

30. $5 + 5 + 5 + 5 =$ how many?

31. Multiplication is a short way of performing what other process? Prove this in the case of six times three.

MULTIPLICATION.

SECTION 8.—1. How much is twice 34?

MODEL. Twice 4 units is 8 units; twice 3 tens is 6 tens. 6 tens and 8 units are 68. *Ans.* 68.

2. How much is 3 times 23? 4 times 120?
3. How much is twice 21? 311 × 3? 101 × 4?
4. How much is twice 44? 22 × 4? 102 × 3?
5. How much is 4 times 1000? 9 times 100?
6. How much is 3 times 2000? 7 times 101?
7. How much is 10 times 7? 10 times 9? 10 times 10? 10 times 100? 10 times 17?
8. How do we multiply a number by 10?
9. How much is 100 times 2? 100 times 9? 100 times 20? 100 times 35? 100 times 68?
10. What easy way is there of multiplying by 100?
11. How much is 20 times 7? 9 × 20? 14 × 20?
12. How much is 200 times 7? 200 times 9?
13. How much is 2000 times 7? 2000 times 9?
14. How much is 4 times 91?

MODEL. 4 times 1 unit is 4 units; 4 times 9 tens is 36 tens, or 3 hundreds and 6 tens. 3 hundreds 6 tens and 4 units are 364. *Ans.* 364.

15. How much is 5 times 61? 4 times 72?
16. How much is twice 640? 8 times 61?
17. If there are 144 steel pens in one box, how many are there in two boxes?
18. How many pages are there in 10 volumes, containing 288 pages each?
19. How many trees has a nurseryman, if he has 100 rows of 17 each?
20. How many pills are there in 200 boxes, each containing 24 pills?

MULTIPLICATION.

SECTION 9.—1. How much is 8 times 24?

MODEL. 8 times 4 units is 32 units, or 3 tens (which we add to the next product) and 2 units. 8 times 2 tens is 16 tens, and 3 tens make 19 tens. 19 tens 2 units are 192. *Ans.* 192.

2. How much is 4 times 43? 32×6? 55×5?
3. How much is 9 times 24? 87×3? 98×2?
4. How much is 7 times 68? 14×8? 79×3?
5. How much is 5 times 103? 84×6? 38×9?
6. How much is 11 times 36? 12 times 51? 9 times 803? 7 times 219? 12 times 45?
7. If a boy learns 13 pages every week, how many pages will he learn in 12 weeks? If he learns 12 pages a week, how many will he learn in 13 weeks?
8. How much is once 13? Twice 13? 3 times 13? 4 times 13? 5 times 13? 6 times 13? 7 times 13? 8 times 13? 9 times 13?
9. How much is 13 times 1? 13 times 2? 13 times 3? 13 times 4? 13 times 5? 13 times 6? 13 times 7? 13 times 8? 13 times 9?
10. How much is once 17? Twice 17? 3 times 17? 4 times 17? 5 times 17? 6 times 17? 7 times 17? 8 times 17? 9 times 17?
11. How much is 17 times 1? 17 times 2? 17 times 3? 17 times 4? 17 times 5? 17 times 6? 17 times 7? 17 times 8? 17 times 9?
12. How much is once 19? Twice 19? 3 times 19? 4 times 19? 5 times 19? 6 times 19? 7 times 19? 8 times 19? 9 times 19?
13. How much is 19 times 1? 19 times 2? 19 times 3? 19 times 4? 19 times 5? 19 times 6? 19 times 7? 19 times 8? 19 times 9?

14. If 7 apples can be bought for a dime, how many can be bought for 17 dimes?

15. A person bought 19 caps, at $3 apiece, and paid for them with 12 five-dollar bills; how many dollars should he receive as change?

16. How much will a boy who earns $5 a week, lack of having earned $100, after working 13 weeks?

17. How many fingers have thirteen persons?

SECTION 10.—Numbers multiplied together are called the **Factors** of their product.

4 times 3 is 12; 4 and 3 are factors of 12.

A number may have more than one set of factors. 4 times 3 is 12; 6 times 2 is 12; 12 has two sets of factors. What are they?

1. What are the factors of 8? Of 15? Of 22? Of 35? Of 26? Of 77? Of 34?

2. Find as many sets of factors as you can for 24. For 30. For 36. For 48. For 16.

3. How much is 12 times 7?

4. ($12 = 4 \times 3$) How much is 4 times 7? 3 times 28?

5. ($12 = 6 \times 2$) How much is 6 times 7? Twice 42?

6. Does it make any difference whether we multiply by a product at once, or by its factors in turn?

7. How much is 18 times 8?

8. ($18 = 9 \times 2$) How much is 9 times 8? Twice 72?

9. ($18 = 3 \times 6$) How much is 3 times 8? 6 times 24?

10. How much is 14 times 13?

MODEL.—14 being 7 times 2, 14 times 13 is equal to 7 times 13, multiplied by 2. 7 times 13 is 91; twice 91 is 182. *Ans.* 182.

11. How much is 18 times 23 ?   15 times 38 ?
12. How much is 16 times 26 ?   14 times 72 ?
13. How much is 21 times 17 ?   35 times 13 ?
14. How much is 25 times 19 ?   32 times 26 ?

NOTE. In multiplying numbers together, choose the easier number to multiply by.

15. If one sheep yields 3 pounds of wool, how many pounds will 45 sheep yield?

MODEL. If 1 sheep yields 3 pounds, 45 sheep will yield 45 times 3 pounds. 45 times 3 equals 3 times 45, or 135. *Ans.* 135 pounds.

16. Twelve make a dozen; how many are 23 dozen?
17. Twenty make a score; how many are 47 score?
18. How many are three score years and ten?
19. Which is the greater, 13 dozen or 7 score?
20. How many dollars will an acre of orange trees produce in a year, if they average $28 a tree, and there are 53 trees to the acre?

MODEL. If one tree produces $28, 53 trees will produce 53 times $28. 53 times $28 = 53 × 4 × 7. Four times 53 is 212; seven times 212 is 1484. *Ans.* $1484.

21. What cost 17 cows, at $56 each?
22. There are 24 hours in one day; how many hours in 23 days? In 29 days?
23. How many trees in four fields, each containing 11 rows, and each row containing 9 trees?
24. How many letters in 37 lines, averaging forty letters to the line?
25. How many yards in 14 pieces of cloth, each containing 37 yards?
26. If 7 boys have each 6 hens, and each hen has 8 chickens, how many chickens have the boys in all?

27. John has 23 marbles; Samuel, 5 times as many. How many has Samuel? How many have both?

28. Levi has 16 cents; Simon has 7 times as many. How many have both?

29. A person bought 6 baskets of fruit, each containing 8 peaches, 9 apples, and 7 pears. How many peaches had he? How many apples? How many pears? How many of all three?

30. How many trees in 13 fields, each containing 6 apple trees and 9 pear trees?

31. Every day a boy earned 90 cents, and spent 65; how many cents had he at the end of six days?

32. If I lay in 200 pounds of butter, and use 13 pounds a week for 15 weeks, how much will remain?

33. Bought 23 cows for $48 each; sold them at $60 apiece. What was the profit on each? On the whole?

34. Sold, at $75 each, 18 lots of land that cost $49 apiece. What was the profit on the whole?

35. A ferryman took 17 passengers across a river for 5 cents each, and then lost 15 cents of what they paid him. How much had he left?

36. How many flowers in 10 bouquets, each containing 5 roses, 9 pinks, and 7 daisies?

37. If I travel 19 miles an hour for 6 hours, and then 23 miles an hour for 5 hours, how far will I go in the whole eleven hours?

38. A and B, travelling toward each other, met in 12 hours. How far apart were their starting-points, if A went 10 miles an hour, and B 12?

39. Find first the sum, then the difference, and then the product, of 10 and 35. Of 11 and 36.

40. A farmer, having 40 acres of land, gave away 15 of them; what were the rest worth, at $9 an acre?

41. A farmer, having 85 acres of land, gave his son and daughter each 25 acres. What was the land that he kept worth, at $20 an acre?

42. Sold 12 suits, at $17 for each coat, $5 for each vest, and $8 for each pair of pants. What was received for the whole?

43. If I buy 6 almanacs at 6 cents apiece, and 5 more at 5 cents apiece, and sell them all at 7 cents apiece, what is my profit?

44. How much is 12 times 7 — 7 times 12?

CHAPTER FOURTH.

DIVISION.

SECTION 11.—**Division** is the process of finding how many times one number is contained in another.

Division is denoted by a short horizontal line between two dots $\div$, placed after the number to be divided. $6 \div 2$ is read, and means, *six divided by two.*

The result, or number obtained by dividing, is called the **Quotient**. $6 \div 2 = 3$; we have divided 6 by 2, and 3 is the quotient.

One number is not always contained in another an exact number of times. Something may be left over, which is called the **Remainder**. $7 \div 2 = 3$, and 1 over; 3 is the quotient, and 1 the remainder.

DIVISION TABLE.

1 in	2 in	3 in	4 in
1, once.	2, once.	3, once.	4, once.
2, twice.	4, twice.	6, twice.	8, twice.
3, 3 times.	6, 3 times.	9, 3 times.	12, 3 times.
4, 4 times.	8, 4 times.	12, 4 times.	16, 4 times.
5, 5 times.	10, 5 times.	15, 5 times.	20, 5 times.
6, 6 times.	12, 6 times.	18, 6 times.	24, 6 times.
7, 7 times.	14, 7 times.	21, 7 times.	28, 7 times.
8, 8 times.	16, 8 times.	24, 8 times.	32, 8 times.
9, 9 times.	18, 9 times.	27, 9 times.	36, 9 times.
10, 10 times.	20, 10 times.	30, 10 times.	40, 10 times.
11, 11 times.	22, 11 times.	33, 11 times.	44, 11 times.
12, 12 times.	24, 12 times.	36, 12 times.	48, 12 times.

5 in	6 in	7 in	8 in
5, once.	6, once.	7, once.	8, once.
10, twice.	12, twice.	14, twice.	16, twice.
15, 3 times.	18, 3 times.	21, 3 times.	24, 3 times.
20, 4 times.	24, 4 times.	28, 4 times.	32, 4 times.
25, 5 times.	30, 5 times.	35, 5 times.	40, 5 times.
30, 6 times.	36, 6 times.	42, 6 times.	48, 6 times.
35, 7 times.	42, 7 times.	49, 7 times.	56, 7 times.
40, 8 times.	48, 8 times.	56, 8 times.	64, 8 times.
45, 9 times.	54, 9 times.	63, 9 times.	72, 9 times.
50, 10 times.	60, 10 times.	70, 10 times.	80, 10 times.
55, 11 times.	66, 11 times.	77, 11 times.	88, 11 times.
60, 12 times.	72, 12 times.	84, 12 times.	96, 12 times.

9 in	10 in	11 in	12 in
9, once.	10, once.	11, once.	12, once.
18, twice.	20, twice.	22, twice.	24, twice.
27, 3 times.	30, 3 times.	33, 3 times.	36, 3 times.
36, 4 times.	40, 4 times.	44, 4 times.	48, 4 times.
45, 5 times.	50, 5 times.	55, 5 times.	60, 5 times.
54, 6 times.	60, 6 times.	66, 6 times.	72, 6 times.
63, 7 times.	70, 7 times.	77, 7 times.	84, 7 times.
72, 8 times.	80, 8 times.	88, 8 times.	96, 8 times.
81, 9 times.	90, 9 times.	99, 9 times.	108, 9 times.
90, 10 times.	100, 10 times.	110, 10 times.	120, 10 times.
99, 11 times.	110, 11 times.	121, 11 times.	132, 11 times.
108, 12 times.	120, 12 times.	132, 12 times.	144, 12 times.

DIVISION.

1. How many times is 4 contained in 28? 5 in 15? 6 in 24? 3 in 27? 2 in 16? 1 in 9? 5 in 35? 6 in 36? 2 in 12? 3 in 36? 5 in 55?

2. How many times is 7 contained in 14? 9 in 36? 10 in 40? 8 in 56? 11 in 33? 9 in 54? 12 in 72? 7 in 70? 10 in 70? 11 in 99? 8 in 40?

3. How many times is 2 contained in 24? 12 in 24? 3 in 18? 6 in 18? 12 in 108? 8 in 96? 9 in 99? 11 in 121? 12 in 144? 10 in 120? 9 in 45?

4. How many times is 3 contained in 26? (*Ans.* 8 *times, and* 2 *over.*) 6 in 43? 9 in 20? 2 in 11? 10 in 95? 9 in 91? 5 in 43? 8 in 66? 4 in 14?

5. What is the quotient, and what the remainder, in the following? $14 \div 5$? $68 \div 7$? $9 \div 2$? $39 \div 4$? $112 \div 11$? $62 \div 10$? $61 \div 7$? $51 \div 8$? $19 \div 2$? $63 \div 5$? $140 \div 12$? $150 \div 12$? $133 \div 11$? $27 \div 10$?

6. If 6 tops cost 30 cents, what will 1 cost?

MODEL. If 6 tops cost 30 cents, 1 top will cost as many cents as 6 is contained times in 30, or 5. *Ans.* 5 cents.

7. If 4 lemons cost 8 cents, what will one cost?

8. If 8 vests cost $32, how much is that apiece?

9. What will 1 primer cost, if 9 cost 81 cents?

10. If 7 hats cost $42, what will one hat cost?

11. If 1 hat cost $6, how many can I buy for $42?

12. $5 make an eagle; how many eagles in $25?

13. At 10 cents a ride, how many rides can a person take for 80 cents? For 100 cents?

14. If 72 marbles are divided equally among 9 boys, how many marbles will each receive?

15. If it takes 48 yards of calico to make 4 dresses, how many yards will it take for 1 dress?

16. How many rows of 12 each will 96 pins make?

17. 120 units make how many dozen?

18. At what rate per hour is a steamboat moving, when it goes 77 miles in 7 hours?

19. How many ten-gallon cans will be required to hold 50 gallons of milk?

20. How many benches, holding 11 children each, will it take to hold 88 children?

SECTION 12.—1. What is the product of 4 and 3? $12 \div 4 =$ how many? $12 \div 3 =$ how many?

2. What is the product of 3 and 7? How many times is 3 contained in 21? 7 in 21?

3. What is the product of 5 and 4? How many times is 5 contained in 20? 4 in 20?

4. When we divide a product of two factors by one of the factors, what do we get?

5. What are the factors of 22? $22 \div 11 =$ how many? $22 \div 2 =$ how many?

6. Divide 7 times 4 by 4. Divide 7 times 4 by 7.

7. How many times is 4×3 contained in 12×11?

8. How many times is $15 - 11$ contained in 13×4?

9. How many times is $4 + 17$ contained in 9×21?

10. How many pigs, at $9 apiece, should a man give in exchange for 9 sheep at $8 apiece?

11. A father buys ten dimes' worth of oranges, at the rate of three for a dime, and divides them equally between his daughter and two sons; how many oranges does each receive?

12. A makes $6 a day for 5 days. B buys some goods for $48, and sells them for $53. How many such lots of goods must B buy and sell, to make as much as A?

13. How many times is 10 contained in 30? In 50? In 110? In 60? In 10? In 70?

14. How do we divide a number by 10? *Ans.* By cutting off its last figure.

15. 10 in 100, how many times? 10 in 140? 10 in 230? 10 in 360? 10 in 780? 10 in 970?

16. If the figure thus cut off is not 0, what do we call it? *Ans.* Remainder; 10 in 75 is contained 7 times, and 5 remainder.

17. 10 in 105, how many times? 10 in 174? 10 in 211? 10 in 863? 10 in 329? 10 in 468?

18. How do we divide a number by 100? *Ans.* By cutting off its last two figures.

19. How many times is 100 contained in 200? In 700? In 1000? In 1300? In 1500? In 3000?

20. If the figures thus cut off are not naughts, what do we call them? *Ans.* Remainder; 100 in 354 is contained 3 times, and 54 remainder.

21. How many times is 100 contained in 109? In 483? In 1007? In 1708? In 1780? In 9657?

22. If 30 loaves are divided equally among 10 poor families, how many loaves will each receive?

23. If $1100 is to be raised in equal parts from 100 persons, how many dollars must each pay?

24. If 3 boys and 7 girls obtain 1200 good marks during a term, what is the average to each?

25. How many hundred-dollar watches can be bought for $1900? For $2500? For $2999?

DIVISION.

SECTION 13.—1. 3 in 6390, how many times?

MODEL. 3 is contained in 6 thousands, 2 thousand times; in 3 hundreds, 1 hundred times; in 9 tens, 3 tens, or thirty times; in 0 units, 0 times. *Ans.* 2130 times.

2. How many times is 4 contained in 844? 5 in 550? 2 in 68? 3 in 936? 4 in 480? 2 in 246?

3. How many times is 9 contained in 819?

MODEL. 9 is not contained in 8. In 81 tens it is contained 9 tens, or 90 times; in 9, once. *Ans.* 91.

4. How many times is 8 contained in 648? 7 in 4277? 6 in 5406? 4 in 3284? 3 in 2793?

5. Find the quotient. 1082÷2. 1869÷3. 4860÷6. 4055÷5. 128÷4. 3690÷9. 2408÷8.

6. How many times is 7 contained in 224?

MODEL. 7 is not contained in 2. 7 in 22 tens, 3 tens, or 30 times, and 1 ten over. 1 ten and 4 units are 14 units; 7 in 14, twice. *Ans.* 32.

7. How many 9's in 207? In 153? In 288? In 414? In 171? In 468? In 243? In 747?

8. 576 is how many times 6? How many times 8? How many times 12? How many times 4?

9. In how many hours can a boat go 192 miles, if it moves at the rate of 8 miles an hour?

10. After sailing 56 miles, how long will it take this boat to perform the rest of its trip of 192 miles?

11. How many dozen oranges can be sold out of a load containing 372 oranges?

12. How many bags, containing 8 pecks each, will be needed to hold 456 pecks of oats?

13. There being 7 days in 1 week, how many weeks are there in 364 days?

14. If a person's expenses amount to $11 a week, how many weeks will $319 last him?

15. A girl is fastening buttons on cards in rows of 12 each. How many rows will she make out of 400 buttons, and how many buttons over?

16. By selling 6 barrels of flour for $69, a man made $3. What did his flour cost him a barrel?

17. If there are 3 sheets used in printing a pamphlet, how many pamphlets will 1086 sheets make?

18. Allowing 5 beets to a bunch, how many bunches will 493 beets make, and how many beets over?

19. A person drew $765 out of the bank in five-dollar bills; how many bills had he? How many of these bills must he use, to pay a debt of $95?

SECTION 14.—1. 2 in 20, how many times? 20 in 20? 2 in 40, how many times? 20 in 40?

2. 3 in 60, how many times? 30 in 60? 3 in 90, how many times? 30 in 90? 30 in 120?

3. 40 in 80, how many times? In 160? In 200? 50 in 200, how many times? In 250? In 300?

4. How do we divide by 40? *Ans.* By cutting off the last figure of the number to be divided, and then dividing by 4, prefixing the remainder, if any, to the figure cut off, for the true remainder.

5. How do we divide by 50? How do we divide by 80? By 90?

6. How do we divide by 60? By 70?

7. How many times is 60 contained in 360? In 480? In 660? In 720? In 2400? In 1200?

8. 70 is contained in 140 how many times? In 490? In 840? In 2100? In 770? In 280?

9. 80 in 80? 80 in 160? 80 in 400? 80 in 720? 80 in 2400? 80 in 480? 80 in 640?

10. 90 in 90? 90 in 270? 90 in 450? 90 in 630? 90 in 810? 90 in 1800? 90 in 3600?

11. How many pounds of meat, at 20 cents a pound, can be bought for 180 cents?

12. How many rows of 30 each will 930 trees make?

13. How many 50-dollar bills make $1000?

14. How many 100-dollar bills make $3000?

15. How many 60-acre farms will 2400 acres make?

16. How many 70-dollar watches will $840 buy?

17. How many mules, at $100 each, can I buy for $1235, and how much will be left over?

18. How many years will it take a certain house to yield me $960, if I let it for $120 a year?

19. Into how many gangs of 40 men can 280 laborers be divided?

20. A drover, having 720 sheep, divides them into flocks of 80 each; how many flocks does he make? After selling 5 of these flocks, how many sheep has he left? What are they worth, at $8 each?

21. In how many days will a man travel 2000 miles, if he travels 100 miles a day? If 200 a day?

22. If a person lays up $200 a year, in how many years will he be worth $4000? $8000?

23. In a certain library are 2400 volumes. How many cases will be needed to hold them, allowing 10 shelves to each case and 30 volumes to each shelf?

24. Divide 1200×2 by $700-400$.

SECTION 15.—1. What is the result of addition called? Of subtraction? Of multiplication? Of division?

2. What operation is performed, to produce a quotient? A sum? A product? A difference?

3. Find first the product, then the quotient, then the sum, and then the difference, of 5 and 105?

4. How many times is the difference between 75 and 105 contained in their sum?

5. How many times is $132 \div 11$ contained in 72×4?

6. A boy caught 25 chub, 56 perch, 13 pout, and 11 eels. If he divided them equally with four of his friends, how many fish had each?

7. From New York to Albany, the distance is 144 miles; from New York to Sing Sing, by the same road, 30 miles. How long will it take to go from Sing Sing to Albany, at the rate of 12 miles an hour?

8. A borrows $238 from B; how much a month must he give B, to pay the debt in 12 months?

9. A farmer had 18 ducks, 12 turkeys, 11 geese, 39 chickens, and 10 guinea-fowls. After selling 34 of the whole number, he put the rest in 4 coops, dividing them equally; how many were in each coop?

10. How many pounds of butter, at 40 cents a pound, should be given for 20 pounds of meat, worth 18 cents a pound?

11. A person invested $960 in apples, at $3 a barrel. How many trees, at the rate of 2 barrels to a tree, did it take to produce these apples?

12. How long will 8000 pounds of flour last 25 men, allowing each man 4 pounds a day?

13. Six boys put in 40 cents each, and bought some melons, at 12 cents apiece; how many did they buy?

14. A person, having to make a journey of 154 miles, travelled 13 miles an hour for 4 hours, and 14 miles an hour for 3 hours. How many hours will it take him to complete his journey, travelling the rest of the way 12 miles an hour?

15. How many boxes, containing 20 pens each, can a person fill from 5 large boxes, holding 144 each?

16. A boy wishes to buy a pony for $100. If he lays up $8 a month for 12 months, how much will he lack?

17. If you owe a person $50 for one bill of goods and $47 for another, how much of the whole debt can you pay with 10-dollar bills?

18. If 11 men can do a job in 33 days, how long will it take 1 man to do it? How long, 3 men?

19. Two men travel from the same place in opposite directions, one 10 miles an hour, the other 6 miles an hour; how far apart are they at the end of 11 hours?

20. A house was bought for $1200, and sold for $1500. The profit was divided between 4 persons; what was the share of each?

21. Laura has 17 chestnuts, twice as many hickory nuts, and as many peanuts as she has chestnuts and hickory nuts. If she divides them all among 6 of her playmates, how many nuts does each receive?

22. Twenty acres of land are bought for $80. How many dollars per acre must they be sold for, that the purchaser may double his money?

CHAPTER FIFTH.

FRACTIONS.

SECTION 16.—1. If we divide a pear into two equal parts, what is each part called? *Ans.* A **half**.

2. How many halves in a whole?

3. How do we get half of anything? *Ans.* By dividing it into 2 equal parts.

4. How do we find half of a number? *Ans.* By dividing it by 2.

5. What is half of 6? Of 10? Of 18? Of 60? Of 2? Of 1? Of 80? Of 100? Of 1000?

 6. If we divide a pear into three equal parts, what is each part called? *Ans.* One **third**.

7. How many thirds in a whole?

8. How do we get a third of anything?

9. How do we find one third of a number? *Ans.* By dividing it by 3.

10. What is one third of 12? Of 15? Of 27? Of 36? Of 42? Of 3? Of 300? Of 1?

11. If we divide a pear into four equal parts, what is each part called? *Ans.* One **fourth**, or **quarter**.

12. How many fourths in a whole?

13. How many quarters in 1?

14. How do we get a fourth of anything?

15. How do we find one fourth, or quarter, of a number? *Ans.* By dividing it by—what?

FRACTIONS.

16. How much is one fourth of 4? Of 1? Of 44? Of 88? Of 96? One quarter of 100? Of 200?

17. If we divide a pear into five equal parts, what is each part called? *Ans.* One **fifth**.

18. How many fifths in a whole, or 1?

19. How do we find one fifth of a number?

20. How much is one fifth of 25? Of 95? Of 125? Of 1? Of 200? Of 300? Of 3000?

21. If we divide a pear into six equal parts, what is each part called? *Ans.* One **sixth**.

22. How many sixths in 1?

23. How do we find one sixth of a number?

24. How much is one sixth of 6? Of 1? Of 72? Of 120? Of 366? Of 480? Of 612?

25. If we divide a whole into seven equal parts, what is each part called? *Ans.* One **seventh**.

26. How many sevenths in 1?

27. How do we find one seventh of a number?

28. How much is one seventh of 63? Of 91? Of 112? Of 140? Of 280? Of 700? Of 350?

29. If we divide a whole into eight equal parts, what is each part called? *Ans.* One **eighth**.

30. If we divide a whole into nine equal parts, what is each part called? *Ans.* One **ninth**.

31. How do we get *tenths, elevenths,* &c.? *Ans.* By dividing a whole into 10, 11, &c., equal parts.

32. How do we get thirteenths? Fifteenths? Thirtieths? Thirty-seconds? Forty-firsts?

33. How many twelfths in 1? How many nineteenths? How many twenty-firsts?

34. How do we find one eighth of a number? How do we find one tenth? One thirteenth? One thirtieth?

35. How much is one eighth of 72? Of 32? Of 96? One ninth of 117? Of 135? One tenth of 20? Of 200? Of 2000? One eleventh of 132? Of 220? Of 1650? One twelfth of 168?

36. How much greater is one twelfth of 600 than one twentieth of 100?

SECTION 17.—1. What are one half, one third, one fourth, &c., called? *Ans.* **Fractions.**

2. Into what is the pear on the left divided? Into what, the pear on the right? Which is greater, one half or one third?

3. Which is greater, one half a line or one fourth of the same line?

---

4. The more parts we divide a thing into, the larger or smaller must those parts be?

5. Walter divides a quart of chestnuts equally among his 3 brothers; what part of a quart will each have? He then divides a quart equally among his 4 sisters; what part will each have? Which will have the larger share, one of the brothers or one of the sisters?

6. Which is greater, one fifth or one sixth?

7. How much is one fifth of 60? One sixth of 60? Which is the greater?

8. If A can dig a certain ditch in 12 days, what part of it can he do in 1 day?

9. If B can dig the same ditch in 10 days, what part can he dig in 1 day? Can he do more or less than A?

10. If a pipe can empty a cistern in 7 hours, what part of the cistern can it empty in one hour?

11. If a person leaves his property, worth $1800, to his wife and 8 children, in equal parts, what fraction of the whole will each have, and how many dollars?

12. Four equal partners make a profit of $2400; what part of the whole, and how much, must each receive?

13. If you put up ten pounds of tea in packages of one pound, what part of the whole do you put in each package?

14. 1 is what part of 10? Of 5? Of 15? Of 9?

15. The earth turns on its axis in 24 hours; what part of a revolution does it perform in 1 hour?

16. There being 100 cents in a dollar, what part of a dollar is 1 cent?

17. If a barrel of flour will last a family 30 days, what part of it will last them one day?

18. One is what part of 9 times 11?

19. How many fifths make 1? How many ninths?

20. Which is greater, 5 fifths or 9 ninths?

21. A person having $85, earned $115 more, and then gave away one fourth of what he had. How much did he give away? How much did he retain?

FRACTIONS.

SECTION 18.—1. If we divide an orange into 4 equal parts, what is one such part called? What are 2 such parts called? *Ans.* Two fourths. What are 3 such parts called?

2. If you divide 25 cents equally among 5 beggars, what part will one beggar receive? Two beggars? Three beggars? Four beggars? What part will all five beggars receive? *Ans.* 5 fifths, or the whole.

3. What part of 10 is 1? What part of 10 is 2?

MODEL. 1 is 1 tenth of 10; and 2 is twice 1 tenth, or 2 tenths. *Ans.* 2 is 2 tenths of 10.

4. What part of 11 is 1? What part of 11 is 3?

5. What part of 12 is 1? What part of 12 is 5?

6. What part of 9 is 7? 8 is what part of 13? 15 is what part of 19? What fraction of 100 is 21?

7. If we divide one by 3 (that is, divide 1 into 3 equal parts), what is the quotient? *Ans.* One third.

8. If we divide 2 by 3, what is the quotient? *Ans.* Two thirds.

9. If we divide 1 by 5, what is the quotient? 2 by 5? 3 by 5? 4 by 5?

10. If we divide 7 by 5, what is the quotient, and what the remainder?

11. What may we call this remainder, 2? *Ans.* As the 2 is to be divided by 5, we may call it 2 fifths.

12. If we divide 11 by 8, what is the quotient? *Ans.* 1 and 3 eighths.

13. How many times is 4 contained in 23? *Ans.* 5 and 3 fourths.

14. (Express the remainder as above.) $53 \div 7 =$ how many? $99 \div 8$? $46 \div 4$? $37 \div 10$? $120 \div 11$?

FRACTIONS.

SECTION 19.—1. What is a Fraction?

Ans. A **Fraction** is one or more of the equal parts into which a whole is divided.

2. Learn how fractions are written.

One half	$\frac{1}{2}$	One tenth.	$\frac{1}{10}$
One third	$\frac{1}{3}$	One eleventh	$\frac{1}{11}$
One fourth	$\frac{1}{4}$	One twelfth	$\frac{1}{12}$
One fifth	$\frac{1}{5}$	Two twelfths	$\frac{2}{12}$
One sixth	$\frac{1}{6}$	Three twelfths	$\frac{3}{12}$
One seventh	$\frac{1}{7}$	Four twenty-firsts	$\frac{4}{21}$
One eighth	$\frac{1}{8}$	Five thirty-seconds	$\frac{5}{32}$
One ninth	$\frac{1}{9}$	Six hundredths	$\frac{6}{100}$

3. A written fraction consists of two numbers, one below the other, with a line between.

The number below the line is called the **Denominator.** It shows into how many equal parts the whole is divided, and gives name to these parts.

The number above the line is called the **Numerator.** It shows how many of the equal parts denoted by the Denominator are taken.

The Numerator and the Denominator, taken together, are called the **Terms** of the fraction.

$\frac{5}{9}$ is a fraction. 5 and 9 are its Terms. 9 is the Denominator, and shows that the whole is divided into *nine* equal parts, making each part one *ninth*. 5 is the Numerator, and shows that *five* of these equal parts are taken. In reading, name the numerator first, and pronounce *ths* distinctly at the end,—*five nin*THS.

4. Read each fraction; name its terms; mention its numerator and denominator; tell what each shows.

$\frac{4}{5}$. $\frac{5}{9}$. $\frac{11}{13}$. $\frac{3}{31}$. $\frac{9}{52}$. $\frac{17}{64}$. $\frac{18}{70}$. $\frac{22}{401}$. $\frac{13}{472}$. $\frac{141}{1000}$.

5. Write in figures:—Two fifths. Three sevenths. Five sixths. Seven elevenths. Five nineteenths.

6. Write in figures:—Seven tenths. Seven hundredths. Seven thousandths. Fifty-one thousandths. Nineteen sixtieths. Four seventy-fifths.

7. How much is $\frac{1}{3}$ of 18? How much is $\frac{2}{3}$ of 18?

MODEL. *One* third of 18 is 6, and *two* thirds are twice 6, or 12. *Ans.* 12.

8. How much is $\frac{3}{4}$ of 12? $\frac{4}{5}$ of 10? $\frac{3}{5}$ of 20?
9. How much is $\frac{5}{6}$ of 18? $\frac{2}{7}$ of 21? $\frac{4}{5}$ of 35?
10. How much is $\frac{9}{10}$ of 100? $\frac{4}{9}$ of 36? $\frac{5}{7}$ of 77?
11. How much is $\frac{3}{11}$ of 121? $\frac{5}{8}$ of 40? $\frac{5}{9}$ of 54?
12. How much is $\frac{5}{12}$ of 144? $\frac{2}{3}$ of 60? $\frac{3}{8}$ of 72?
13. How much is $\frac{7}{20}$ of 160? $\frac{5}{9}$ of 45? $\frac{7}{8}$ of 88?
14. How much is $\frac{1}{11}$ of 90? $\frac{2}{5}$ of 50? $\frac{3}{7}$ of 91?
15. How much is $\frac{13}{40}$ of 40? $\frac{7}{12}$ of 96? $\frac{6}{11}$ of 99?
16. How much greater is $\frac{7}{10}$ of 80 than $\frac{3}{6}$ of 48?
17. How much less is $\frac{2}{5}$ of 35 than $\frac{5}{11}$ of 33?
18. Which is greater, $\frac{9}{12}$ of 108 or $\frac{4}{9}$ of 63?

19. A person having $1550 divided it into five equal parts, and gave one of these parts to each of his three children. What part of the whole did the children receive, and how much?

20. A road 160 miles long is composed of 10 equal sections. What fraction of the whole road are 6 of these sections, and how many miles do they contain?

21. Two brothers are equal heirs with 6 other persons to an estate of $8000. What part of the whole do the two brothers receive, and how much money?

22. If the rent of a house for 30 days is $90, how much will it be for 17 days?

FRACTIONS.

SECTION 20.—1. What is a Fraction?

2. What does *one third* mean? *Ans.* One of three equal parts into which a whole is divided.

3. What does *two thirds* mean? *Ans.* Two of three equal parts into which a whole is divided.

4. What does one fourth mean? Two fourths? $\frac{3}{4}$?

5. How is this expression, $6\frac{3}{4}$, read? *Ans.* Six and three fourths.

6. Of what does it consist? *Ans.* Of a whole number (6), and a fraction ($\frac{3}{4}$).

7. What is a number that consists of a whole number and a fraction called? *Ans.* A **Mixed Number.**

8. How many fourths in 6?

MODEL. In 1 there are 4 fourths, and in 6 six times 4 fourths, or 24 fourths. *Ans.* $\frac{24}{4}$.

9. How many fourths in $6\frac{3}{4}$?

MODEL. In 1 there are 4 fourths, and in 6 six times 4 fourths, or 24 fourths. 24 fourths and 3 fourths are 27 fourths. *Ans.* $\frac{27}{4}$.

10. How many eighths in 7? In $7\frac{3}{8}$? In $9\frac{5}{8}$?
11. How many fifths in 11? In $11\frac{1}{5}$? In $12\frac{3}{5}$?
12. How many ninths in 12? In $12\frac{4}{9}$? In $11\frac{2}{9}$?
13. How many sevenths in 4? In $4\frac{3}{7}$? In $8\frac{5}{7}$?
14. How many tenths in 10? In $10\frac{1}{10}$? In $4\frac{3}{10}$?
15. How many twelfths in 6? In $5\frac{1}{12}$? In $8\frac{7}{12}$?
16. 5 = how many halves? $5\frac{1}{2}$? $6\frac{1}{2}$? $8\frac{1}{2}$?
17. How many thirds in 3? In 7? In 9? In 5?
18. $8\frac{1}{6}$ = how many sixths? $9\frac{5}{6}$? $8\frac{3}{6}$? $4\frac{2}{6}$? 4 = how many elevenths? $4\frac{2}{11}$? $7\frac{3}{11}$? $9\frac{1}{11}$? $6\frac{7}{11}$?

19. How many halves in $50\frac{1}{2}$? Twentieths in $5\frac{11}{20}$? Hundredths in $2\frac{1}{100}$? Fortieths in $4\frac{3}{40}$? Thirds in $13\frac{1}{3}$? Eighths in $14\frac{3}{8}$? Quarters in $7\frac{3}{4}$?

20. What kind of fractions are $\frac{3}{2}$, $\frac{4}{3}$, &c.? *Ans.* **Improper Fractions.**

21. What is an Improper Fraction? *Ans.* A Fraction whose numerator is equal to, or greater than, its denominator.

22. What is a fraction called, whose numerator is less than its denominator? *Ans.* **A Proper Fraction.**

SECTION 21.—1. How many thirds do $4\frac{1}{3}$ equal?

2. When you say $\frac{13}{3}$ for $4\frac{1}{3}$, do you change the *form* of the mixed number? Do you change its *value?*

3. What is the process of changing the form, without changing the value, called? *Ans.* **Reduction.**

4. In the last Section, what kind of numbers did you reduce? To what kind of fractions did you reduce them?

5. 5 units equal how many thirds? 15 thirds, then, equal how many units?

6. How many units in 16 thirds?

MODEL. Since 3 thirds make 1 unit, 16 thirds make as many units as 3 thirds are contained times in 16 thirds, or $5\frac{1}{3}$. *Ans.* $5\frac{1}{3}$.

7. How many units in $\frac{16}{2}$? In $\frac{18}{4}$? In $\frac{18}{8}$?
8. How many units in $\frac{17}{3}$? In $\frac{19}{5}$? In $\frac{25}{9}$?
9. How many units in $\frac{42}{5}$? In $\frac{42}{6}$? In $\frac{37}{4}$?
10. How many wholes in $\frac{18}{2}$? In $\frac{20}{5}$? In $\frac{30}{10}$?
11. How many wholes in $\frac{36}{12}$? In $\frac{37}{11}$? In $\frac{63}{12}$?
12. Which is greater, $\frac{69}{12}$ or $\frac{57}{11}$? $\frac{27}{3}$ or $\frac{47}{6}$?
13. Which is greater, $\frac{64}{8}$ or $\frac{32}{4}$? $\frac{101}{10}$ or $\frac{72}{8}$?
14. How many times 1 in $\frac{43}{5}$? In $\frac{201}{10}$?
15. How many times 1 is $\frac{150}{3}$? $\frac{141}{7}$? $\frac{145}{12}$?

16. $\frac{48}{26}$=how many units? $\frac{23}{12}$? $\frac{32}{11}$? $\frac{85}{4}$?

17. Which is greater, $\frac{55}{4}$ or $\frac{55}{5}$? $\frac{62}{7}$ or $\frac{62}{8}$?

18. How many dollars in $\frac{29}{4}$ of a dollar? In 29 quarter-dollars? In 29 half-dollars?

19. In this Section, what kind of fractions have you reduced?

20. To what have you reduced them?

SECTION 22.—1. What is a Fraction? A Proper Fraction? An Improper Fraction?

2. What is a Mixed Number?

3. What is meant by the Reduction of Fractions?

4. How many halves in 12? How many units in $\frac{24}{3}$?

5. How do you reduce units to halves? How do you reduce halves to units?

6. How do you reduce thirds to units? Units to thirds?

7. How do you reduce units to sixths? How do you reduce sixths to units?

8. How do you reduce ninths to units? Units to ninths?

9. Reduce 8 to twentieths. Reduce $\frac{160}{8}$ to units.

10. Reduce $\frac{17}{6}$ to a mixed number. Reduce $2\frac{5}{6}$ to an improper fraction.

11. At a dinner, 47 plates of pie were called for; how many pies were used, if each plate contained quarter of a pie?

12. A druggist is putting up soda in powders of $\frac{1}{4}$ of a dram each. If he uses $9\frac{1}{4}$ drams of soda, how many powders does he put up?

13. How many bows, requiring $\frac{1}{6}$ of a yard of ribbon each, can be made out of $7\frac{5}{6}$ yards?

14. At a lecture 43 dollars, 24 half-dollars, and 32 quarter-dollars, were taken in. How much in all was received?

15. A pint is one eighth of a gallon; how many pints in $13\frac{3}{8}$ gallons? How many gallons in 141 pints?

16. If a boy can saw and split $\frac{1}{12}$ of a cord of wood in a day, how many such boys will it take to saw and split $4\frac{7}{12}$ cords in a day?

17. How many cords would the boy just mentioned saw and split in 300 days?

18. How many dollars will 53 tops cost, at the rate of $\frac{1}{20}$ of a dollar apiece?

19. How many quarter-pound weights will it take to balance 56 pounds? How many half-pound weights?

20. If an apple costs $\frac{1}{100}$ of a dollar, how many apples can you buy for $1? For $2? For $5?

21. How many lots $\frac{1}{16}$ of an acre in size can be laid out from 5 acres? From 7 acres? From $2\frac{3}{16}$ acres?

22. How many inches in $2\frac{1}{8}$ of an inch?

23. One day is $\frac{1}{7}$ of a week; how many weeks in 50 days?

24. How much is $\frac{2}{3}$ of 18? 14 is how many times 2?

25. $\frac{2}{3}$ of 18 is how many times 2?

26. $\frac{3}{5}$ of 60 is how many times 9?

27. $\frac{4}{7}$ of 56 is how many times 8?

28. $\frac{4}{5}$ of 50 is how many times $15+5$?

29. $\frac{3}{8}$ of 72 is how many times $17-8$?

30. $\frac{2}{3}$ of 84 is how many times $\frac{1}{2}$ of 16?

31. $\frac{1}{4}$ of 96 is how many times $\frac{2}{3}$ of 12?

FRACTIONS.

SECTION 23.—1. 6 is $\frac{1}{3}$ of what number?

MODEL. If 6 is *one* third of the required number, *three* thirds, or the whole, must be three times 6, or 18. *Ans.* 6 is $\frac{1}{3}$ of 18.

2. 4 is $\frac{1}{8}$ of what number?
3. 7 is $\frac{1}{16}$ of what number?
4. 9 is $\frac{1}{6}$ of what number?
5. 3 times 4 is $\frac{1}{4}$ of what number?
6. The sum of 5 and 9 is $\frac{1}{4}$ of what number?
7. 2×8 is $\frac{1}{6}$ of what number?
8. $\frac{1}{3}$ of 21 is $\frac{1}{6}$ of what number?
9. $\frac{2}{3}$ of 21 is $\frac{1}{2}$ of what number?
10. $\frac{1}{5}$ of 25 is $\frac{1}{10}$ of what number?
11. $\frac{3}{5}$ of 25 is $\frac{1}{3}$ of what number?
12. $\frac{3}{9}$ of 45 is $\frac{1}{5}$ of what number?
13. $2 + 8 - 3$ is $\frac{1}{4}$ of what number?
14. 10 is $\frac{1}{11}$ of what number? 10 is $\frac{1}{12}$ of what?
15. If a man can do $\frac{1}{10}$ of a piece of work in 2 hours, how long will it take him to do the whole?

MODEL. If a man can do *one* tenth of a piece of work in 2 hours, to do *ten* tenths, or the whole, will require 10 times 2 hours, or 20 hours. *Ans.* 20 hours.

16. If a family consume $\frac{1}{12}$ of a barrel of flour in 11 days, how long will the barrel last them?
17. If a boat is 3 hours in performing $\frac{1}{4}$ of its trip, how long, at that rate, will its whole trip take?
18. If a locomotive can go six miles in $\frac{1}{6}$ of an hour, how many miles can it go in one hour?
19. With $24 collected from one customer, $16 from another, and $11 from a third, a person paid $\frac{1}{2}$ of his taxes; what did his taxes amount to?
20. $7 + 8 + 5 + 4$ is $\frac{1}{3}$ of what number?

5

21. 6 is $\frac{2}{3}$ of what number?

MODEL. If 6 is *two* thirds of the required number, *one* third of it is $\frac{1}{2}$ of 6, or 3; and *three* thirds, or the whole, are 3 times 3, or 9. *Ans.* 6 is $\frac{2}{3}$ of 9.

22. 4 is $\frac{2}{5}$ of what number?
23. 12 is $\frac{3}{4}$ of what number?
24. 36 is $\frac{6}{11}$ of what number?
25. 48 is $\frac{4}{9}$ of what number?
26. 27 is $\frac{3}{10}$ of what number?
27. 54 is $\frac{9}{13}$ of what number?
28. 33 is $\frac{3}{4}$ of what number?
29. 40 is $\frac{5}{8}$ of what number?
30. 3 times 10 is $\frac{5}{8}$ of what number?
31. 12—7 is $\frac{5}{8}$ of what number?
32. $\frac{1}{6}$ of 72 is $\frac{4}{5}$ of what number?
33. $\frac{3}{4}$ of 44 is $\frac{11}{12}$ of what number?
34. $\frac{2}{5}$ of 105 is $\frac{3}{7}$ of what number?
35. 8 is $\frac{2}{7}$ of what? 28 is how many times 4?
36. 8 is $\frac{2}{7}$ of how many times 4?
37. 9 is $\frac{3}{10}$ of how many times 5?

MODEL. If 9 is *three* tenths of a certain number, *one* tenth of it is $\frac{1}{3}$ of 9, or 3; and *ten* tenths, or the whole, are 10 times 3, or 30. 30 is 6 times 5. *Ans.* 6 times.

38. 14 is $\frac{7}{12}$ of how many times 8?
39. 64 is $\frac{8}{9}$ of how many times 12?
40. $\frac{1}{2}$ of 36 is $\frac{3}{4}$ of how many times 2?
41. Charles lost 18 marbles, which were $\frac{6}{13}$ of all he had. How many had he at first?
42. A boy, having received $5 apiece from 6 persons, had only $\frac{6}{10}$ of what he needed to buy a pony. What was the price of the pony?

FRACTIONS.

43. 70 is $\frac{7}{9}$ of how many times 5?
44. 70 is $\frac{7}{9}$ of how many times $\frac{1}{2}$ of 10?
45. 35 is $\frac{5}{6}$ of how many times 3?
46. 35 is $\frac{5}{6}$ of how many times $\frac{1}{4}$ of 12?
47. 12 is $\frac{3}{14}$ of how many times $\frac{2}{3}$ of 12?
48. 16 is $\frac{8}{17}$ of how many times $\frac{1}{6}$ of 18?
49. $\frac{1}{4}$ of 80 is how many times 5?
50. $\frac{1}{4}$ of 80 is how many times $\frac{1}{6}$ of 30?
51. $\frac{4}{11}$ of 44 is how many times $\frac{1}{12}$ of 96?
52. $\frac{3}{50}$ of 100 is how many times 6?
53. $\frac{7}{9}$ of 36 is how many times $\frac{1}{6}$ of 84?
54. $\frac{3}{7}$ of 49 is how many times $\frac{1}{6}$ of 27?
55. A is 80 years old, and $\frac{5}{16}$ of his age is $\frac{1}{2}$ of B's; how old is B?
56. From Buffalo to Syracuse is 150 miles, and $\frac{2}{5}$ of this distance is $\frac{1}{2}$ the distance from Buffalo to Verona; how far is it from Buffalo to Verona?
57. Wilmington is 28 miles from Philadelphia, and $\frac{1}{4}$ of this distance is $\frac{1}{14}$ of the distance from Philadelphia to Baltimore. How far is Philadelphia from Baltimore?
58. From Boston to Newburyport is 36 miles, which is 8 miles more than $\frac{1}{2}$ the distance from Boston to Portsmouth. How far is Portsmouth from Boston?
59. A man, having bought some corn for $80, sold it for $\frac{4}{5}$ of its cost; did he gain or lose, and how much?
60. A man bought some corn for $80, which was $\frac{4}{5}$ of what he sold it for; did he gain or lose, and how much?
61. A farmer had 24 cows, and $\frac{3}{4}$ as many sheep. Had he more sheep or cows, and how many more?

FRACTIONS.

SECTION 24.—1. What is meant by Reducing a fraction?

2. If we divide a pie into two equal parts, each part is called $\frac{1}{2}$. If we divide each half into two equal parts, we get four equal parts in all, and each is $\frac{1}{4}$ of the whole.

It is clear that two of these fourths equal one half,—or that $\frac{2}{4}$ *may be reduced to* $\frac{1}{2}$.

3. $\frac{2}{4} = \frac{1}{2}$. What operation performed on $\frac{2}{4}$ gives $\frac{1}{2}$? *Ans.* Dividing its terms by 2. $\quad \frac{2 \div 2}{4 \div 2} = \frac{1}{2}$

4. $\frac{1}{2} = \frac{2}{4}$. What operation performed on $\frac{1}{2}$ gives $\frac{2}{4}$? *Ans.* Multiplying its terms by 2. $\quad \frac{1 \times 2}{2 \times 2} = \frac{2}{4}$

5. What principle may we lay down? *Ans.* The value of a fraction is not changed by dividing or multiplying its terms by the same number.

6. When we divide both terms of a fraction by 2, show why we do not change its value. *Ans.* We get only *half* as many parts as before, but each part is *twice* as large.

7. When we multiply both terms of a fraction by 2, show why we do not change its value. *Ans.* We get *twice* as many parts as before, but each part is only *half* as large.

8. When is a fraction *in its lowest terms?* *Ans.* When no number greater than 1 is exactly contained in both terms. $\frac{1}{2}$ is in its lowest terms; $\frac{2}{4}$ is not, because 2 is exactly contained in its numerator and denominator.

9. Is $\frac{2}{3}$ in its lowest terms? $\frac{3}{4}$? $\frac{3}{6}$? $\frac{5}{9}$? $\frac{5}{10}$?

REDUCTION OF FRACTIONS. 53

10. How is a fraction reduced to its lowest terms? *Ans.* By dividing its terms by whatever number or numbers, greater than 1, are exactly contained in both.

11. Reduce $\frac{3}{27}$ to its lowest terms. *Ans.* $\frac{1}{9}$.

12. Reduce the following to their lowest terms:—
$\frac{2}{4}$. $\frac{5}{10}$. $\frac{7}{21}$. $\frac{4}{6}$. $\frac{10}{18}$. $\frac{28}{35}$. $\frac{15}{24}$. $\frac{33}{55}$. $\frac{9}{12}$. $\frac{14}{49}$.

13. Reduce $\frac{75}{100}$ to its lowest terms.

MODEL. Dividing both terms by 5, we reduce the fraction to $\frac{15}{20}$. Again dividing both terms by 5, we get $\frac{3}{4}$. *Ans.* $\frac{3}{4}$.

14. Reduce the following to their lowest terms:—
$\frac{60}{90}$. $\frac{42}{56}$. $\frac{45}{75}$. $\frac{80}{112}$. $\frac{75}{105}$. $\frac{21}{84}$. $\frac{42}{128}$. $\frac{96}{108}$.

15. Reduce $\frac{36}{60}$ to its lowest terms. $\frac{40}{56}$. $\frac{200}{300}$.

16. Reduce $\frac{22}{99}$ to its lowest terms. $\frac{22}{88}$. $\frac{120}{360}$.

17. How many halves in $1\frac{8}{4}$? In $5\frac{3}{6}$? In $\frac{75}{10}$?

18. How many thirds in $7\frac{8}{12}$? In $\frac{37}{21}$? In $\frac{40}{6}$?

19. How many fourths in $3\frac{4}{8}$? How many halves?

20. How many fifths in $7\frac{4}{10}$? In $3\frac{6}{10}$? In $5\frac{8}{10}$?

21. Reduce $\frac{88}{12}$ to a mixed number.

NOTE. Always see that a fraction occurring in an answer is in its lowest terms.

22. Reduce $\frac{305}{10}$ to a mixed number. $\frac{104}{12}$. $\frac{110}{8}$.

23. How much is $\frac{24}{32}$ of 16?

MODEL. $\frac{24}{32} = \frac{3}{4}$. $\frac{1}{4}$ of 16 is 4, and $\frac{3}{4}$ is 3 times 4, or 12. *Ans.* 12.

NOTE. It is often best to reduce a fraction to its lowest terms before operating with it.

24. How much is $\frac{12}{18}$ of 6? $\frac{20}{24}$ of 12?

25. How much is $\frac{66}{72}$ of 24? $\frac{36}{52}$ of 20?

26. 10 is $\frac{20}{24}$ of what number?

27. $\frac{3}{8}$ of 16 is $\frac{24}{21}$ of what number?

28. 24 is $\frac{30}{65}$ of how many times 2?

SECTION 25.—1. What may we do to the terms of a fraction, without changing its value?

2. How is a fraction reduced to its lowest terms?

3. How may a fraction be reduced to higher terms? *Ans.* By multiplying its terms by the same number; $\frac{1}{2}=\frac{2}{4}$.

4. Explain why, when we multiply both terms of a fraction by 3, we do not change its value.

5. How many sixths in $\frac{2}{3}$?

MODEL. In 1 there are 6 sixths, and in $\frac{2}{3}$ there are $\frac{2}{3}$ of 6 sixths, or 4 sixths. *Ans.* $\frac{4}{6}$.

6. How many ninths in $\frac{1}{3}$? How many tenths in $\frac{3}{5}$?

7. How many 24ths in $\frac{1}{6}$? In $\frac{5}{8}$? In $\frac{7}{12}$? In $\frac{3}{4}$?

8. How many 36ths in $\frac{4}{9}$? In $\frac{2}{3}$? In $\frac{3}{4}$? In $\frac{5}{12}$?

9. How many 18ths in $\frac{5}{6}$? In $\frac{4}{3}$? In $1\frac{1}{3}$? In $1\frac{2}{9}$?

10. Reduce $1\frac{3}{10}$ to twentieths. To 30ths. To 50ths.

11. Reduce $3\frac{5}{6}$ to twelfths. Reduce $2\frac{5}{8}$ to 63rds.

12. How many twelfths in $\frac{1}{4}$? In $\frac{1}{2}$? In $\frac{1}{6}$? In $\frac{1}{3}$?

13. What do you observe with respect to the fractions $\frac{3}{12}$, $\frac{6}{12}$, $\frac{2}{12}$, and $\frac{4}{12}$? *Ans.* They have *a common denominator* (12).

14. Reduce $\frac{1}{2}$ and $\frac{3}{5}$ to fractions that have a common denominator.

MODEL. $2 \times 5 = 10$; the common denominator is 10. $\frac{1}{2}$ equals $\frac{5}{10}$; $\frac{3}{5}$ equals $\frac{6}{10}$. *Ans.* $\frac{5}{10}$, $\frac{6}{10}$.

15. Reduce to fractions with a common denominator $\frac{1}{3}$ and $\frac{1}{7}$; $\frac{3}{4}$ and $\frac{5}{8}$; $\frac{2}{3}$ and $\frac{7}{10}$; $\frac{1}{2}$ and $\frac{5}{6}$.

16. Reduce to fractions with a common denominator $2\frac{1}{4}$ (that is, $\frac{9}{4}$) and $\frac{2}{3}$; $\frac{1}{6}$ and $1\frac{1}{7}$ ($\frac{8}{7}$); $1\frac{1}{2}$ and $1\frac{1}{3}$.

17. Reduce to fractions having a common denominator $\frac{1}{2}$, $\frac{1}{3}$, and $\frac{1}{5}$; $\frac{4}{5}$, $\frac{2}{3}$, and $\frac{1}{4}$; $\frac{1}{10}$, $\frac{5}{7}$, and $\frac{1}{2}$.

REDUCTION OF FRACTIONS.

18. If an ounce is $\frac{1}{12}$ of a pound, how many ounces in $\frac{1}{2}$ of a pound? In $\frac{5}{6}$ of a pound?

19. One cent is one hundredth of a dollar; what fraction of a dollar is 10 cents? 25 cents? 75 cents?

20. How many inches, or 36ths of a yard, in $\frac{1}{6}$ of a yard? In $\frac{2}{3}$ of a yard? In $\frac{3}{4}$ of a yard?

21. How many eighths of a gallon in a gallon and a half? In 5 gallons? In quarter of a gallon?

22. Reduce $\frac{1}{2}$, $\frac{3}{4}$, and $\frac{7}{8}$, to fractions with *the least* common denominator.

MODEL. As 8, the third denominator, exactly contains the others (2 and 4), it is the least common denominator. $\frac{1}{2}$ equals $\frac{4}{8}$; $\frac{3}{4}$ equals $\frac{6}{8}$. *Ans.* $\frac{4}{8}$, $\frac{6}{8}$, $\frac{7}{8}$.

23. Reduce to fractions with *the least* common denominator $\frac{5}{7}$ and $\frac{1}{28}$. $\frac{5}{6}$, $\frac{1}{2}$, and $\frac{2}{3}$. $\frac{1}{4}$, $\frac{5}{16}$, and $\frac{3}{8}$.

24. Reduce $\frac{7}{8}$, $\frac{2}{3}$, and $1\frac{3}{4}$, to fractions having the least common denominator. Reduce $\frac{1}{20}$, $\frac{2}{5}$, $\frac{1}{4}$, and $\frac{7}{10}$.

25. Reduce $\frac{1}{6}$, $\frac{5}{12}$, $\frac{1}{3}$, and $\frac{3}{4}$, to fractions with the least common denominator. Reduce $\frac{5}{18}$, $\frac{5}{36}$, $\frac{5}{6}$, $\frac{5}{4}$.

26. Reduce $\frac{5}{18}$, $\frac{2}{5}$, $1\frac{1}{3}$ (that is, $\frac{4}{3}$), to fractions with the least common denominator. Reduce $1\frac{1}{6}$, $\frac{7}{30}$, $1\frac{2}{5}$.

27. Reduce $\frac{3}{4}$ and $\frac{5}{6}$ to fractions having a common denominator.

28. Reduce $\frac{3}{4}$ and $\frac{5}{6}$ to fractions having *the least* common denominator.

MODEL. 2, being a factor of both denominators 4 and 6, may be rejected from their product. $4 \times 6 = 24$. $24 \div 2 = 12$, *least common den.* $\frac{3}{4}$ equals $\frac{9}{12}$; $\frac{5}{6}$ equals $\frac{10}{12}$. *Ans.* $\frac{9}{12}$, $\frac{10}{12}$.

29. Reduce $\frac{1}{9}$ and $\frac{5}{6}$ to fractions having the least common denominator. Reduce $\frac{3}{10}$ and $\frac{3}{4}$. Reduce $\frac{5}{6}$ and $\frac{3}{8}$. Reduce $1\frac{1}{12}$ and $\frac{5}{8}$. Reduce $\frac{9}{10}$ and $\frac{5}{6}$.

SECTION 26.—1. What is Addition? By what sign is it denoted?

2. What is the result of addition called?
3. How much are 9 times 5 and $\frac{1}{5}$ of 5?
4. How much are 6 times 7 and $\frac{1}{6}$ of 42?
5. How much are 4 times 6 and $\frac{1}{3}$ of 33?
6. How many are 2 apples and 3 apples? 2 books and 3 books? 2 ninths and 3 ninths?
7. How many are 4 tops and 5 tops? 4 tenths and 5 tenths? 4 elevenths and 5 elevenths?
8. How much are 5 thirds and 6 thirds? $\frac{5}{3}+\frac{6}{3}$?
9. How much are 7 sixths and 8 sixths? $\frac{7}{6}+\frac{8}{6}$?
10. What is the sum of $\frac{6}{8}$ and $\frac{2}{8}$? $\frac{3}{13}$ and $\frac{5}{13}$?
11. What is the sum of $\frac{11}{14}$ and $\frac{5}{14}$?

MODEL. 11 fourteenths and 5 fourteenths are $\frac{16}{14}$, or $1\frac{2}{14}$, equal to $1\frac{1}{7}$. *Ans.* $1\frac{1}{7}$.

12. What is the sum of $\frac{7}{8}$ and $\frac{3}{8}$? Of $\frac{5}{8}$ and $\frac{7}{8}$?
13. What is the sum of $\frac{7}{12}$, $\frac{5}{12}$, and $\frac{6}{12}$?
14. Add $\frac{9}{16}$, $\frac{7}{16}$, and $\frac{12}{16}$. Add $\frac{7}{6}$, $\frac{8}{6}$, and $\frac{2}{6}$.
15. How much is $\frac{1}{5}+\frac{2}{5}+\frac{3}{5}$? How much is $\frac{2}{7}+\frac{3}{7}$?
16. How much is $\frac{1}{4}+\frac{5}{4}+\frac{3}{4}+\frac{7}{4}$?
17. How much is 7 increased by $\frac{3}{10}+\frac{17}{10}$?
18. How much is 3 added to the sum of $\frac{4}{9}$ and $1\frac{1}{9}$?
19. Add 4 and $\frac{1}{7}$. Add 4 and $\frac{1}{7}+\frac{5}{7}$.
20. Add 4, 2, and the sum of $\frac{4}{6}$ and $\frac{5}{6}$.
21. How much are $4\frac{4}{7}$ and $2\frac{4}{7}$? $3\frac{1}{5}$ and $4\frac{4}{5}$?
22. How much are $5\frac{2}{3}$ and $3\frac{1}{3}$? $1\frac{4}{5}$ and $7\frac{1}{5}$?
23. What is the sum of $\frac{1}{2}$ of 26 and $\frac{1}{2}$ of 18?
24. What is the sum of $\frac{1}{12}$, $5\frac{5}{12}$, and $2\frac{8}{12}$?

MODEL. $\frac{1}{12}$, $\frac{5}{12}$, and $\frac{8}{12}$, are $\frac{14}{12}$, equal to $1\frac{2}{12}$, or $1\frac{1}{6}$. 1 and 5 and 2 are 8. *Ans.* $8\frac{1}{6}$.

25. How much is $2\frac{1}{6}+\frac{5}{6}+3\frac{4}{6}$? $1\frac{3}{10}+\frac{9}{10}+5$?

26. How much are $1\frac{5}{8}$, $\frac{7}{8}$, and $\frac{1}{8}$ of 19?

27. A person who owned $\frac{7}{18}$ of a steamboat, bought from two other parties $\frac{3}{18}$ and $\frac{5}{18}$; what part of the boat had he then?

28. A man, having a farm of 253 acres, divides it equally among his 5 children. If his eldest child already had $25\frac{3}{5}$ acres, how much land has he now?

29. Reduce $\frac{3}{4}$ and $\frac{5}{6}$ to fractions having a common denominator.

SECTION 27.

1. How much are $\frac{3}{4}$ and $\frac{5}{4}$?

2. How much are $\frac{3}{4}$ and $\frac{5}{6}$?

MODEL. $\frac{3}{4}=\frac{9}{12}$; $\frac{5}{6}=\frac{10}{12}$. $\frac{9}{12}+\frac{10}{12}=\frac{19}{12}$, or $1\frac{7}{12}$. Ans. $1\frac{7}{12}$.

3. What is the difference between Examples 1 and 2?

4. How do we add fractions that have a common denominator?

5. What must we first do, when they have not a common denominator?

6. What is the sum of $\frac{1}{2}$ and $\frac{1}{3}$? Of $\frac{1}{2}$ and $\frac{1}{4}$?

7. What is the sum of $\frac{1}{4}$ and $\frac{1}{6}$? Of $\frac{1}{3}$ and $\frac{1}{6}$?

8. What is the sum of $\frac{1}{4}$ and $\frac{1}{5}$? Of $\frac{1}{4}$ and $\frac{1}{8}$?

9. What is the sum of $\frac{1}{6}$ and $\frac{1}{5}$? Of $\frac{1}{6}$ and $\frac{1}{8}$?

10. How much is $\frac{1}{4}+\frac{1}{8}$? How much is $\frac{1}{4}+\frac{1}{6}$?

11. How much is $\frac{2}{3}+\frac{1}{2}$? How much is $\frac{3}{4}+\frac{1}{3}$?

12. How much is $\frac{5}{6}+\frac{3}{8}$? How much is $\frac{2}{3}+\frac{2}{5}$?

13. How much is $\frac{2}{3}+\frac{5}{6}$? How much is $\frac{4}{5}+\frac{7}{8}$?

14. $\frac{3}{10}+\frac{5}{8}$= how much? $\frac{5}{12}+\frac{1}{4}$= how much?

15. $\frac{1}{6}+\frac{2}{3}+\frac{1}{4}$= how much? $\frac{1}{2}+\frac{1}{3}+\frac{5}{6}$= how much?

16. $\frac{1}{3}+\frac{1}{8}+\frac{1}{6}=$ how much? $\frac{3}{8}+\frac{1}{2}+\frac{5}{10}=$ how much?

17. How much is $1+\frac{2}{5}+\frac{1}{4}$? How much is $\frac{2}{8}+1\frac{1}{6}$?

18. How much is $1+2+\frac{3}{8}+\frac{1}{16}$? $1\frac{3}{8}+2\frac{1}{16}$?

19. How much is $4\frac{3}{7}+2\frac{1}{2}$? How much is $5\frac{2}{11}+3\frac{1}{3}$?

20. How much is $1\frac{5}{24}+7\frac{7}{12}$? $3\frac{3}{40}+4\frac{5}{8}$?

21. How much are $\frac{1}{4}$ of 19 and $\frac{1}{6}$ of 18?

22. How much are $\frac{1}{4}$ of 23 and $\frac{1}{6}$ of 42?

23. How much are $\frac{1}{10}$ of 20, $\frac{1}{6}$ of 18, and $\frac{1}{5}$ of 10?

24. A market-woman sold $\frac{1}{5}$ of her eggs for 40 cents, $\frac{1}{15}$ of them for 9 cents, and $\frac{3}{5}$ of them for 57 cents. What part of her eggs did she sell, and for how much?

25. A can do $\frac{1}{8}$ of a piece of work in 1 day, B $\frac{1}{4}$, and C $\frac{1}{6}$. How much can all three do in a day?

26. If a boy who had $4\frac{1}{4}$, earned $2\frac{3}{8}$ more, and had $1\frac{1}{10}$ given him, how many dollars had he then?

27. A walked half a mile in $\frac{1}{6}$ of an hour, $\frac{3}{8}$ of a mile in $\frac{7}{30}$ of an hour, and $1\frac{3}{4}$ miles in $\frac{8}{15}$ of an hour. How many miles did he walk, and how long did it take him?

28. If I buy some muslin for $3\frac{1}{5}$, lace for $2\frac{3}{10}$, calico for $2\frac{1}{20}$, and $\frac{1}{2}$ dozen collars for $7\frac{9}{20}$, how much change should I receive for a twenty-dollar bill?

29. If a person buys a razor for $\frac{3}{4}$ of a dollar and a strop for $\frac{5}{8}$ of a dollar, for how much must he sell them both in order to make half a dollar?

30. Helen had $3\frac{7}{25}$, Louise $5\frac{28}{50}$, and Mary $1\frac{1}{5}$. They put their money together, and divided it equally among several poor persons, so that each received $\frac{1}{5}$ of the sum. How many poor persons were there, and how much did each get?

SUBTRACTION OF FRACTIONS.

SECTION 28.—1. What is Subtraction? By what sign is it denoted?

2. What is the result of subtraction called?
3. How much is 11 times $3 - \frac{1}{8}$ of 40?
4. How much is 9 times $8 - \frac{2}{3}$ of 36?
5. How much is 7 times $4 - \frac{3}{4}$ of twice 8?
6. 4 pins − 3 pins = how many pins? 4 ninths − 3 ninths = how many ninths? $\frac{4}{8} - \frac{3}{8}$?
7. 10 knives − 3 knives = how many knives? 10 elevenths − 3 elevenths? $\frac{10}{11} - \frac{3}{11}$?
8. How much is $\frac{11}{12} - \frac{1}{12}$? How much is $\frac{9}{16} - \frac{3}{16}$?
9. How much is $\frac{21}{10} - \frac{7}{10}$? How much is $2\frac{1}{10} - \frac{7}{10}$?
10. From the sum of $\frac{9}{32}$ and $\frac{21}{32}$ take $\frac{6}{32}$.
11. From the sum of $\frac{3}{40}$, $\frac{17}{40}$, and $\frac{23}{40}$, take $\frac{19}{40}$.
12. From the sum of $\frac{5}{24}$ and $\frac{17}{24}$ take $\frac{11}{24} - \frac{3}{24}$.
13. $\frac{5}{6}$ from $\frac{13}{6}$ leaves how much? $\frac{5}{6}$ from $2\frac{1}{6}$?
14. $\frac{7}{8}$ from $\frac{19}{8}$ leaves how much? $\frac{7}{8}$ from $2\frac{3}{8}$?
15. Reduce $\frac{3}{8}$ and $\frac{4}{8}$ to fractions having a common denominator.

SECTION 29.—1. From $\frac{7}{8}$ take $\frac{5}{8}$.

2. From $\frac{7}{8}$ take $\frac{5}{6}$.

MODEL. $\frac{7}{8}$ equals $\frac{21}{24}$; $\frac{5}{6}$ equals $\frac{20}{24}$. $\frac{21}{24} - \frac{20}{24} = \frac{1}{24}$. *Ans.* $\frac{1}{24}$.

3. What is the difference between Ex. 1 and 2?
4. How do we subtract one fraction from another, when they have a common denominator?
5. What must we first do, when they have not a common denominator?
6. From $\frac{1}{2}$ take $\frac{1}{4}$. Take $\frac{1}{4}$ from $\frac{1}{2}$.
7. From $\frac{1}{8}$ take $\frac{1}{8}$. Take $\frac{1}{8}$ from $\frac{1}{4}$.

FRACTIONS.

8. How much is $\frac{2}{3}-\frac{1}{8}$? How much is $\frac{3}{4}-\frac{1}{6}$?
9. How much is $\frac{8}{9}-\frac{2}{3}$? How much is $\frac{7}{8}-\frac{3}{5}$?
10. How much is $\frac{19}{20}-\frac{3}{10}$? How much is $\frac{17}{24}-\frac{5}{8}$?
11. How much is $1\frac{17}{18}-\frac{5}{12}$? How much is $2\frac{31}{36}-\frac{4}{9}$?
12. How much is $3\frac{9}{16}-\frac{5}{24}$? How much is $4\frac{15}{18}-\frac{1}{4}$?
13. How much is $\frac{5}{3}-\frac{1}{2}$? How much is $1\frac{1}{3}-\frac{1}{2}$?
14. How much is $\frac{13}{10}-\frac{2}{5}$? How much is $1\frac{3}{10}-\frac{2}{5}$?
15. How much is $2\frac{1}{6}-\frac{3}{4}$? How much is $5\frac{3}{7}-\frac{11}{14}$?
16. How much is $4\frac{3}{8}-\frac{15}{16}$? How much is $3\frac{1}{2}-\frac{5}{6}$?
17. How much is $6\frac{1}{4}-\frac{2}{5}$? How much is $2\frac{1}{6}-\frac{7}{9}$?
18. How much is $\frac{3}{4}-\frac{1}{9}$? $2\frac{3}{4}-\frac{1}{9}$? $2\frac{3}{4}-1\frac{1}{9}$?
19. How much is $\frac{7}{10}-\frac{1}{5}$? $4\frac{7}{10}-\frac{1}{5}$? $4\frac{7}{10}-3\frac{1}{5}$?
20. How much is $5\frac{7}{8}-\frac{2}{7}$? $4\frac{19}{20}-\frac{5}{6}$? $3\frac{2}{15}-\frac{1}{10}$?
21. How much is $3\frac{2}{3}-1\frac{1}{4}$? $5\frac{1}{2}-2\frac{17}{40}$? $4\frac{5}{6}-\frac{1}{18}$?
22. What remains, if we take $\frac{2}{13}$ from 5?

NOTE. We reduce one of the 5 units to thirteenths, and then subtract. $5=4+1$, or $4\frac{13}{13}$. $\frac{2}{13}$ from $4\frac{13}{13}$ leaves $4\frac{11}{13}$. *Ans.* $4\frac{11}{13}$.

23. From 3 take $\frac{1}{4}$. Take $\frac{5}{6}$ from 7. $\frac{1}{2}$ from 9.
24. From 5 take $\frac{2}{3}$. Take $\frac{3}{4}$ from 6. $\frac{2}{3}$ from 1.
25. From 8 take $\frac{5}{7}$. Take $\frac{5}{12}$ from 7. $\frac{3}{14}$ from 1.
26. From $6\frac{1}{5}$ take $1\frac{3}{4}$.

NOTE. $\frac{1}{5}=\frac{4}{20}$; $\frac{3}{4}=\frac{15}{20}$. As we can not take $\frac{15}{20}$ from $\frac{4}{20}$, we reduce one of the 6 units to $\frac{20}{20}$, and add it to $\frac{4}{20}$, making $\frac{24}{20}$. $1\frac{15}{20}$ from $5\frac{24}{20}$ leaves $4\frac{9}{20}$. *Ans.* $4\frac{9}{20}$.

27. From $2\frac{1}{3}$ take $1\frac{3}{4}$. Subtract $3\frac{5}{6}$ from $5\frac{2}{7}$.
28. From $3\frac{1}{4}$ take $1\frac{5}{6}$. Subtract $4\frac{13}{20}$ from $9\frac{1}{4}$.
29. From $9\frac{1}{4}$ take $2\frac{3}{4}$. Subtract $5\frac{4}{7}$ from 20.
30. From 50 take $40\frac{3}{11}$. Take $99\frac{1}{2}$ from 100.
31. From the sum of $\frac{4}{5}$ and $\frac{7}{12}$ take $\frac{5}{6}$.
32. From $\frac{1}{5}$ of 28 take the sum of $1\frac{1}{4}$ and $\frac{2}{3}$.

SUBTRACTION OF FRACTIONS.

SECTION 30.—1. A grocer, having a bushel of potatoes, sold $\frac{1}{4}$ of it to one customer, $\frac{1}{8}$ to another, and $\frac{3}{10}$ to a third. What part remained unsold?

2. From a farm of 100 acres were taken three fields, containing $2\frac{1}{3}$, $4\frac{1}{6}$, and $3\frac{7}{12}$ acres. How many acres were left?

3. If the sum of $\frac{1}{3}$ and $\frac{2}{15}$ of a person's age is 14 years, how old is he?

4. From two remnants of calico, containing respectively $4\frac{3}{4}$ and $8\frac{7}{8}$ yards, were cut $12\frac{11}{16}$ yards for a dress. How much did what was left lack of 1 yard?

5. The difference between $\frac{2}{3}$ and $\frac{1}{4}$ of F's age was 26 years; how old was he?

6. A stage, after making $\frac{1}{4}$ and $\frac{2}{5}$ of its trip, had 7 miles yet to go; how long was its trip?

MODEL. $\frac{1}{4}=\frac{5}{20}$; $\frac{2}{5}=\frac{8}{20}$; $\frac{5}{20}+\frac{8}{20}=\frac{13}{20}$. The whole trip was $\frac{20}{20}$ of itself; when the stage had made $\frac{13}{20}$ of the trip, there remained $\frac{20}{20}-\frac{13}{20}$, or $\frac{7}{20}$. If $\frac{7}{20}$ of the trip was 7 miles, $\frac{1}{20}$ was $\frac{1}{7}$ of 7 miles, or 1 mile; and $\frac{20}{20}$, or the whole trip, was 20 times 1 mile, or 20 miles. *Ans.* 20 miles.

7. A man performed $\frac{1}{2}$ of his journey in the morning, $\frac{1}{6}$ of it in the afternoon, and the rest (12 miles) in the evening. How long was the journey?

8. A person divided $101 equally between his two daughters. The elder then spent 25\frac{1}{4}$, and gave away 15\frac{3}{8}$; how much of her share had she left?

9. Two thirds of a certain rod is blue, $\frac{2}{5}$ of it red, and the rest white. If the white part is 3 inches long, how long is the whole rod?

10. How old am I, if the difference between $\frac{1}{8}$ and $\frac{1}{6}$ of my age is 3 years?

11. Bought some paper for 12\frac{18}{100}$; sold it for 13\frac{1}{20}$. By how much did the cost exceed the profit?

12. Three pans contained respectively 2$\frac{1}{3}$, 4$\frac{7}{8}$, and 3$\frac{3}{4}$ quarts of milk. If 1$\frac{1}{6}$ quarts were spilled, how many quarts were left?

13. John and Cyrus had 6 dozen eggs each; John sold 2$\frac{1}{2}$ dozen, Cyrus 3$\frac{1}{4}$ dozen. How many dozen did Cyrus have left, and how many eggs less than John?

14. A can do a piece of work in 12 days, and B in 8 days; what part can each do in one day? How much can both together do in one day? After they have worked one day, how much of the job will remain?

15. C can do a piece of work in 4 days, and D in 6 days. How much can both, working together, do in one day, and how much will then remain to be done?

16. E can do a certain job in 9 hours, and F in 6 hours. After they have worked together at it one hour, what part of the job remains to be done?

SECTION 31.—1. What is Multiplication? By what sign is it denoted?

2. What is the result of multiplication called?

3. 5 times 3 equals what? 5 times 3 apples equal what? 5 times 3 sevenths? 5 times 3 ninths?

4. How much is 5 times $\frac{3}{8}$?* 5 times $\frac{3}{8}$?

5. How much is 4 times $\frac{1}{2}$? 6 times $\frac{2}{3}$?

6. How much is 7 times $\frac{3}{4}$? 3 times $\frac{2}{7}$?

7. How much is 9 times $\frac{3}{5}$? 8 times $\frac{4}{9}$?

---

* Reduce the products to whole or mixed numbers, and fractions in the answers to their lowest terms.

MULTIPLICATION OF FRACTIONS. 63

8. Multiply $\frac{3}{4}$ by 2.

NOTE. To multiply $\frac{3}{4}$ by 2, we may double the *number* of parts. Twice *three* fourths is *six* fourths. *Ans.* $\frac{6}{4}$, or $\frac{3}{2}$.

Or, we may double the *size* of the parts. *Halves* are twice as great as *fourths* (as we found on p. 52). Hence, twice three *fourths* is three *halves*. *Ans.* $\frac{3}{2}$. The answers agree.

In the first case, we multiply the numerator by 2: $\frac{3 \times 2}{4} = \frac{6}{4}$.

In the second case, we divide the denominator by 2: $\frac{3}{4 \div 2} = \frac{3}{2}$.

The second method is shorter, when it can be used, because it brings the answer in its lowest terms at once.

9. How much is $\frac{3}{10} \times 5$? $\frac{3}{10} \times 2$? $\frac{5}{6} \times 3$?
10. Multiply $\frac{5}{16}$ by 4. $\frac{7}{24}$ by 8. $\frac{1}{18}$ by 9.
11. Multiply $\frac{2}{27}$ by 3. $\frac{11}{30}$ by 10. $\frac{50}{64}$ by 8.
12. What is the product of $\frac{7}{48}$ and 6? $\frac{5}{48}$ and 12?
13. What is the product of $\frac{11}{66}$ and 5? $\frac{3}{55}$ and 11?
14. What is the product of $\frac{4 3}{4 2}$ and 9? $\frac{2 7}{2 4}$ and 8?
15. What is the product of 10 and $\frac{3}{6}$? $\frac{4}{14}$ and 7?
16. What two ways are there of multiplying a fraction by a whole number?
17. Which is preferable? Why?
18. What is 4 times $\frac{1}{4}$? 4 times $\frac{3}{4}$? 4 times $\frac{5}{4}$?
19. When you multiply a fraction by its own denominator, what do you get?
20. How much is 5 times $\frac{4}{5}$? 10 times $\frac{7}{10}$?
21. How much is 9 times $\frac{2}{9}$? 19 times $\frac{54}{19}$?
22. How much is 100 times $\frac{7}{100}$? 25 times $\frac{18}{25}$?
23. How much does 4 times $\frac{11}{4}$ lack of 10?
24. What must be added to 15 times $\frac{13}{30}$, to make 7?
25. Which is greater, 5 times $\frac{2}{15}$ or $2 - 1\frac{1}{3}$?
26. 7 times $\frac{10}{7}$ is $\frac{5}{6}$ of what number?
27. 8 times $\frac{6}{16}$ is $\frac{1}{4}$ of what number?
28. 3 times $\frac{8}{27}$ is $\frac{2}{6}$ of what number?

SECTION 32.

1. How much is 5 times $3\frac{4}{9}$?

MODEL. 5 times $\frac{4}{9}$ is $\frac{20}{9}$, or $2\frac{2}{9}$; 5 times 3 is 15. $15 + 2\frac{2}{9} = 17\frac{2}{9}$. Ans. $17\frac{2}{9}$.

2. How much is 6 times $2\frac{1}{3}$? 4 times $8\frac{5}{6}$?
3. How much is 3 times $5\frac{2}{3}$? 7 times $1\frac{5}{14}$?
4. How much is 9 times $4\frac{7}{9}$? 5 times $2\frac{17}{30}$?
5. How much is 10 times $7\frac{3}{4}$? 12 times $3\frac{3}{8}$?
6. Multiply $6\frac{2}{3}$ by 2. By 5. By 7. By 10.
7. What cost 8 vests, at $5\frac{3}{4}$ each?
8. How much ale in 6 cans, holding $2\frac{5}{9}$ pints each?
9. What will 9 hens weigh, averaging $2\frac{3}{8}$ pounds?
10. Multiply 6 by $\frac{1}{3}$.

NOTE. 6 multiplied by $\frac{1}{3}$ must be $\frac{1}{3}$ as much as 6 multiplied by 1,—that is, $\frac{1}{3}$ of 6. Multiplying a number by $\frac{1}{3}$, $\frac{1}{4}$, &c., is therefore equivalent to taking $\frac{1}{3}$, $\frac{1}{4}$, &c., of that number.

11. Multiply 24 by $\frac{1}{4}$. By $\frac{3}{4}$. By $\frac{1}{8}$. By $\frac{3}{8}$.
12. Multiply 18 by $\frac{1}{6}$. By $\frac{5}{6}$. By $\frac{2}{3}$. By $\frac{2}{3}$.
13. How much is $\frac{1}{3}$ of 2?

MODEL. $\frac{1}{3}$ of 1 is $\frac{1}{3}$, and $\frac{1}{3}$ of 2 is twice $\frac{1}{3}$, or $\frac{2}{3}$. Ans. $\frac{2}{3}$.

14. How much is $\frac{1}{7}$ of 2? $\frac{1}{8}$ of 2? $\frac{1}{10}$ of 7?
15. How much is $\frac{1}{4}$ of 3? $\frac{1}{8}$ of 5? $\frac{1}{12}$ of 6?
16. Multiply 6 by $\frac{1}{18}$. Multiply 4 by $\frac{1}{18}$. 7 by $\frac{1}{21}$.
17. How much is $\frac{5}{6}$ of 21?

MODEL. One sixth of 21 is $3\frac{1}{2}$, and five sixths are 5 times $3\frac{1}{2}$, or $17\frac{1}{2}$. Ans. $17\frac{1}{2}$.

18. How much is $\frac{3}{4}$ of 5? $\frac{5}{6}$ of 9? $\frac{7}{8}$ of 11?
19. How much is $\frac{2}{3}$ of 8? $\frac{3}{7}$ of 9? $\frac{5}{12}$ of 14?
20. How much is $\frac{1}{9}$ of 2? $\frac{5}{6}$ of 2? $\frac{7}{9}$ of 2?
21. How much is $\frac{7}{8}$ of 3? $\frac{9}{10}$ of 4? $\frac{4}{15}$ of 16?
22. How much is $\frac{1}{4}$ of 100? $\frac{1}{3}$ of 100? $\frac{2}{3}$ of 100?
23. How much is 7 times $2\frac{3}{4}$? $\frac{5}{8}$ of 4? $\frac{4}{7}$ of 11?

24. How far will a locomotive, moving at the rate of $20\frac{7}{12}$ miles an hour, go in 5 hours? In 8 hours?

25. What cost two pieces of meat, weighing respectively $8\frac{3}{8}$ and $6\frac{19}{8}$ pounds, at $20\frac{1}{8}$ cents a pound?

26. At $\$8\frac{3}{4}$ an acre, what will be the cost of 3 fields containing $3\frac{2}{3}$ acres each?

27. What is the weight of four cheeses, if two of them weigh $9\frac{1}{4}$ pounds each, and the other two $10\frac{7}{18}$ pounds each?

28. If I divide $2 equally among 8 beggars, what part of $2 will each receive? What part of $1?

29. A person bought 56 pounds of butter, and sold $\frac{5}{8}$ of it; how many pounds remained on hand?

30. Emma is $3\frac{3}{4}$ years old; Laura is 5 times as old as Emma. In how many years will Laura be 21?

SECTION 33.—1. How much is $5\frac{2}{3}$ times 7?

MODEL. 5 times 7 is 35; $\frac{2}{3}$ of 7 is $\frac{14}{3}$, or $4\frac{2}{3}$; $35 + 4\frac{2}{3} = 39\frac{2}{3}$. Ans. $39\frac{2}{3}$.

2. How much is $2\frac{1}{2}$ times 24? $3\frac{2}{3}$ times 14?

3. How much is $5\frac{5}{8}$ times 10? $4\frac{3}{4}$ times 15?

4. How much is $8\frac{2}{7}$ times 12? $6\frac{3}{8}$ times 16?

5. How much is $3\frac{1}{4}$ times $\frac{5}{6}$ of 18?

6. What cost $7\frac{5}{8}$ acres, at $10 an acre?

7. If $\frac{3}{4}$ of a ton of hay cost $15, what will 1 ton cost? What will $\frac{7}{8}$ of a ton cost?

8. If $\frac{5}{8}$ of a ton of hay cost $20, what will $\frac{2}{3}$ of a ton cost? What will $2\frac{2}{3}$ tons cost?

9. What are $3\frac{1}{8}$ gallons of petroleum worth, if $\frac{2}{3}$ of a gallon is worth 21 cents?

10. $\frac{1}{2}$ of 4 equals how many? $\frac{1}{2}$ of 4 apples? $\frac{1}{2}$ of 4 ninths? $\frac{1}{2}$ of $\frac{4}{7}$? $\frac{1}{2}$ of $\frac{4}{13}$?

11. $\frac{1}{3}$ of 6 equals how many? $\frac{1}{3}$ of 6 knives? $\frac{1}{3}$ of 6 sevenths? $\frac{1}{3}$ of $\frac{6}{8}$? $\frac{1}{3}$ of $\frac{6}{11}$?

12. How much is $\frac{1}{4}$ of $\frac{8}{9}$? $\frac{1}{5}$ of $\frac{10}{11}$? $\frac{1}{10}$ of $\frac{20}{99}$? $\frac{1}{8}$ of $\frac{72}{91}$? $\frac{1}{20}$ of $\frac{40}{87}$? $\frac{1}{100}$ of $\frac{500}{801}$?

13. What is multiplying by a fraction equivalent to? *Ans.* To taking such a part as is denoted by the fraction. Multiplying by $\frac{1}{6}$ is equivalent to taking $\frac{1}{6}$.

14. Multiply $\frac{12}{13}$ by $\frac{1}{6}$. Multiply $\frac{22}{27}$ by $\frac{1}{11}$. Multiply $\frac{14}{19}$ by $\frac{1}{7}$. Multiply $\frac{60}{77}$ by $\frac{1}{30}$.

15. How much is $\frac{3}{4}$ of $\frac{16}{9}$?

MODEL. *One* fourth of $\frac{16}{9}$ is $\frac{4}{9}$, and *three* fourths are 3 times $\frac{4}{9}$, or $\frac{4}{3}$,—equal to $1\frac{1}{3}$. *Ans.* $1\frac{1}{3}$.

16. How much is $\frac{5}{6}$ of $\frac{12}{25}$? How much is $\frac{2}{3}$ of $\frac{18}{17}$?

17. How much is $\frac{7}{9}$ of $\frac{27}{14}$? How much is $\frac{7}{9}$ of $1\frac{13}{14}$?

18. How much is $\frac{4}{5}$ of $\frac{20}{9}$? How much is $\frac{4}{5}$ of $2\frac{2}{9}$?

19. How much is $\frac{3}{7}$ of $\frac{28}{9}$? How much is $\frac{1}{4}$ of $1\frac{1}{3}$?

20. How much is $\frac{1}{4}$ of $\frac{3}{7}$ of $\frac{28}{9}$?

21. How much is $\frac{2}{3}$ of $1\frac{2}{7}$? How much is $\frac{5}{6}$ of $\frac{6}{7}$?

22. How much is $\frac{5}{6}$ of $\frac{2}{3}$ of $1\frac{2}{7}$?

23. How much is $\frac{2}{9}$ of $\frac{2}{5}$ of $2\frac{13}{16}$?

24. If a person bought $\frac{6}{7}$ of a ship, and then sold $\frac{1}{3}$ of his share, what part of the ship did he retain?

25. When cheese is $\frac{9}{50}$ of a dollar a pound, how much will $\frac{2}{3}$ of a pound cost?

26. Jane's age is $\frac{3}{4}$ of Clara's, and Clara's is $\frac{5}{6}$ of Lucy's. If Lucy is 18, how many years old is Jane?

27. If a cow eats $\frac{1}{12}$ of a ton of hay in a week, what will her daily supply of hay cost when hay is \$21 a ton?

DIVISION OF FRACTIONS.

SECTION 34.—1. What is Division? By what sign is it denoted?

2. What is the result of division called?

3. Dividing by 2 is equivalent to taking what part?

4. Dividing by 3 is equivalent to what? Dividing by 4? Dividing by 7?

5. How many times is 6 contained in 18? How much is $\frac{1}{6}$ of 18? Divide 18 by 6.

6. How many times is 6 contained in 18 sevenths? How much is $\frac{1}{6}$ of $\frac{18}{7}$? Divide $\frac{18}{7}$ by 6.

7. Divide $\frac{8}{2}$ by 2.

NOTE. To divide $\frac{8}{2}$ by 2, we may take $\frac{1}{2}$ of the number of parts; $\frac{1}{2}$ of *eight* halves is *four* halves, or 2. *Ans.* 2.

Or, we may make each part half as great. *Fourths* are half as great as *halves* (as we found on p. 52). Hence, $\frac{1}{2}$ of 8 *halves* is 8 *fourths*, or 2. *Ans.* 2. The answers agree.

In the first case, we divide the numerator by 2: $\frac{8}{2} \div 2 = \frac{4}{2} = 2$

In the second case, we multiply the denominator by 2: $\frac{8}{2} \times 2 = \frac{8}{4} = 2$

The first method is preferable when the whole number is exactly contained in the numerator of the fraction.

8. What two ways are there of dividing a fraction by a whole number? Which is preferable?

9. When can the first method be used?

10. How many times is 6 contained in $\frac{2}{7}$?

NOTE. As 6 is not exactly contained in 2, we have to use the second method: $\frac{2}{7} \times 6 = \frac{2}{42}$, or $\frac{1}{21}$. *Ans.* $\frac{1}{21}$.

11. Divide $\frac{2}{5}$ by 9. Divide $\frac{5}{8}$ by 6. Divide $\frac{7}{5}$ by 8.

12. How many times is 10 contained in $\frac{8}{11}$? In $\frac{12}{13}$?

13. How much is $\frac{1}{4}$ of $\frac{6}{5}$? $\frac{1}{3}$ of $\frac{7}{10}$? $\frac{1}{5}$ of $\frac{8}{15}$?

14. How much is $\frac{3}{4}$ of $\frac{2}{5}$?

MODEL. *One* fourth of $\frac{2}{5}$ is $\frac{2}{20}$, or $\frac{1}{10}$; and *three* fourths are 3 times $\frac{1}{10}$, or $\frac{3}{10}$. *Ans.* $\frac{3}{10}$.

15. How much is $\frac{5}{6}$ of $\frac{3}{5}$? $\frac{7}{8}$ of $\frac{4}{5}$? $\frac{2}{3}$ of $\frac{4}{7}$?

16. How much is $\frac{1}{3}$ of $1\frac{1}{8}$? $\frac{1}{5}$ of $1\frac{2}{3}$? $\frac{2}{5}$ of $1\frac{2}{3}$?

17. How much is $\frac{1}{4}$ of $\frac{6}{5}$? $\frac{1}{4}$ of $1\frac{1}{5}$? $\frac{3}{4}$ of $1\frac{1}{5}$?

18. How much is $\frac{2}{3}$ of $2\frac{3}{4}$? $\frac{4}{5}$ of $2\frac{1}{6}$? $\frac{5}{9}$ of $3\frac{1}{3}$?

19. A, B, and C, own a ferry, having equal shares. B sells $\frac{4}{5}$ of his share to C; what part of the ferry does B own after this sale, and what part C? How much more has C than B?

20. If 1 pound of butter costs $\frac{2}{5}$ of a dollar, how much will $\frac{3}{4}$ of a pound cost?

21. Mary has $$1\frac{4}{10}$ and George $$2\frac{3}{5}$. They divide $\frac{1}{8}$ of what they both have equally between two poor persons. What part of a dollar does each receive?

22. By how much does $\frac{2}{3}$ of $\frac{9}{10}$ exceed $\frac{1}{3}$ of $\frac{3}{4}$?

23. How much is $3\frac{2}{3}$ times $2\frac{3}{4}$?

MODEL. $2\frac{3}{4} = \frac{11}{4}$. 3 times $\frac{11}{4} = \frac{33}{4}$, or $8\frac{1}{4}$. *One* third of $\frac{11}{4}$ is $\frac{11}{12}$, and *two* thirds are twice $\frac{11}{12}$, or $\frac{11}{6} = 1\frac{5}{6}$. $8\frac{1}{4} + 1\frac{5}{6} = 10\frac{1}{12}$. *Ans.* $10\frac{1}{12}$.

24. How much is $2\frac{1}{4}$ times $1\frac{1}{2}$? $1\frac{2}{5}$ times $4\frac{1}{3}$?

25. How much is $4\frac{1}{3}$ times $60\frac{1}{4}$? $3\frac{3}{4}$ times $41\frac{2}{3}$?

26. How much is $5\frac{3}{4}$ times $10\frac{1}{10}$? $6\frac{1}{2}$ times $18\frac{1}{18}$?

SECTION 35.—1. How many halves are there in 1? In 2? In 5?

2. How many times is $\frac{1}{2}$ contained in 1? In 2?

3. Dividing by $\frac{1}{2}$ is equivalent to what? *Ans.* Dividing by $\frac{1}{2}$ is multiplying by 2.

4. Multiplying by $\frac{1}{2}$ is equivalent to what? *Ans.* Multiplying by $\frac{1}{2}$ is taking $\frac{1}{2}$, or dividing by 2.

5. If we divide 8 by $\frac{1}{2}$, what is the quotient? If we multiply 8 by $\frac{1}{2}$, what is the product?

DIVISION OF FRACTIONS.

6. Dividing by $\frac{1}{3}$ is equivalent to what? *Ans.* Dividing by $\frac{1}{3}$ is multiplying by 3.
7. Dividing by $\frac{1}{5}$ is equivalent to what? By $\frac{1}{4}$?
8. How many times is $\frac{1}{4}$ contained in 3? In 9?
9. How many times is $\frac{1}{10}$ contained in 2? In $3\frac{3}{10}$? In $4\frac{9}{10}$? In $6\frac{2}{5}$? In $7\frac{1}{2}$?
10. How many times $\frac{1}{8}$ is $3\frac{3}{8}$? $4\frac{1}{2}$? $5\frac{3}{4}$? $6\frac{2}{3}$?
11. How many times $\frac{1}{9}$ is $\frac{2}{3}$? Divide $\frac{2}{3}$ by $\frac{1}{9}$.
12. How many times $\frac{1}{4}$ is $\frac{3}{4}$? Divide $\frac{3}{4}$ by $\frac{1}{4}$.
13. How many times $\frac{1}{6}$ is $\frac{9}{7}$?
14. How many times $\frac{1}{12}$ is $1\frac{5}{8}$?
15. How many times is $\frac{1}{10}$ contained in $1\frac{1}{15}$?
16. How many times is $\frac{1}{13}$ contained in $2\frac{1}{7}$?
17. How many times is $\frac{5}{6}$ contained in $\frac{3}{4}$?

MODEL. *One* sixth is contained in $\frac{3}{4}$, $\frac{18}{4}$ or $\frac{9}{2}$ times; and *five* sixths, being 5 times as great, is contained in it $\frac{1}{5}$ of $\frac{9}{2}$ times, or $\frac{9}{10}$ times. *Ans.* $\frac{9}{10}$ times.

18. How many times is $\frac{2}{7}$ contained in $\frac{3}{5}$? $\frac{3}{8}$ in $\frac{4}{5}$?
19. How many times is $\frac{3}{8}$ contained in $\frac{4}{5}$? $\frac{4}{5}$ in $\frac{8}{9}$?
20. $\frac{5}{9}$ in $\frac{11}{6}$ how many times? $\frac{5}{6}$ in $1\frac{5}{6}$?
21. $\frac{5}{8}$ in $\frac{25}{16}$ how many times? $\frac{5}{8}$ in $1\frac{9}{16}$?
22. $\frac{7}{12}$ in $4\frac{2}{3}$ how many times? $\frac{3}{11}$ in $1\frac{5}{22}$?
23. $\frac{5}{8}$ in $\frac{8}{9}$ how many times? $1\frac{2}{3}$ in $\frac{8}{9}$?
24. $\frac{7}{4}$ in $\frac{7}{4}$ how many times? $1\frac{3}{4}$ in $1\frac{1}{8}$?
25. $\frac{11}{6}$ in $\frac{22}{7}$ how many times? $1\frac{5}{6}$ in $3\frac{1}{7}$?
26. $3\frac{1}{2}$ in $2\frac{1}{8}$ how many times? $4\frac{2}{3}$ in $5\frac{2}{5}$?
27. How far will a person, walking at the rate of $2\frac{5}{8}$ miles an hour, walk in $3\frac{1}{4}$ hours?
28. If the current of a river moves $2\frac{5}{8}$ miles in $3\frac{1}{3}$ hours, how far will it move in 1 hour?
29. How much is $5\frac{1}{2} \times 2\frac{2}{3}$? How much is $5\frac{1}{2} \div 2\frac{2}{3}$?

FRACTIONS.

SECTION 36.—1. How many times is 9 contained in $100\frac{2}{7}$?

Model. 9 is contained in 100, 11 times and 1 over. 9 is contained in $1\frac{2}{7}$, or $\frac{9}{7}$, $\frac{1}{7}$ time. *Ans.* $11\frac{1}{7}$ times.

2. How many times is 11 contained in $84\frac{1}{2}$?
3. How many times is 7 contained in $42\frac{4\cdot8}{49}$?
4. How many times is 12 contained in $39\frac{3}{7}$?
5. How many times is 5 contained in $43\frac{2}{7}$?
6. How many times is 4 contained in $49\frac{3}{9}$?
7. How many times are 7 apples contained in 14 apples? 7 cups in 14 cups? 7 ninths in 14 ninths?
8. How many times are 5 pins contained in 45 pins? 5 eighths in 45 eighths? $\frac{5}{8}$ in $\frac{45}{8}$? $\frac{5}{7}$ in $\frac{45}{7}$?
9. How do we divide one fraction by another, when they have a common denominator? *Ans.* Divide numerator by numerator, rejecting the denominators.
10. $\frac{40}{9} \div \frac{8}{9} =$ how many? $\frac{72}{7} \div \frac{9}{7}$? $1\frac{3}{5} \div \frac{4}{5}$? $5\frac{5}{8} \div 1\frac{1}{8}$? $4\frac{4}{9} \div 1\frac{2}{9}$? $3\frac{5}{12} \div 1\frac{8}{12}$?
11. What cost $\frac{5}{8}$ of a barrel of cider, at $\$8\frac{1}{2}$ a bar.?
12. A person laid out $\$55\frac{1}{2}$ for fuel. He bought 5 tons of coal at $\$7\frac{1}{4}$ a ton, and spent the rest for wood at $\$3\frac{1}{2}$ a load; how many loads of wood did he buy?
13. If a horse goes $2\frac{7}{8}$ miles in an hour, what part of an hour will it take him to go one mile?
14. A farmer sold $\frac{2}{3}$ of his flock, and had 17 sheep left; how many sheep had he originally?
15. A farmer sold 14 sheep, and had $\frac{2}{3}$ of his flock left; how many sheep had he left?
16. What will a pile of $16\frac{1}{2}$ bushels of potatoes bring, if put up in bags holding $2\frac{2}{3}$ bushels each, and sold for $\$1\frac{9}{10}$ per bag?

FRACTIONS.

17. 7 is $\frac{2}{3}$ of what number?

MODEL. If 7 is *two* thirds of a certain number, *one* third of it is $\frac{1}{2}$ of 7, or $\frac{7}{2}$; and *three* thirds, or the whole, are 3 times $\frac{7}{2}$, or $\frac{21}{2}$, which equals $10\frac{1}{2}$. *Ans.* $10\frac{1}{2}$.

18. 9 is $\frac{4}{5}$ of what number?
19. 12 is $\frac{4}{7}$ of what number?
20. $\frac{2}{5}$ of 20 is $\frac{4}{7}$ of what number?
21. $\frac{9}{10}$ of 30 is $\frac{3}{4}$ of how many times 8?
22. $\frac{7}{8}$ of 32 is $\frac{3}{5}$ of how many times $\frac{1}{3}$?
23. $\frac{5}{6}$ of 16 is $\frac{3}{4}$ of how many times 10?
24. $\frac{2}{7}$ of 25 is $\frac{5}{8}$ of how many times 12?
25. $\frac{8}{9}$ of 63 is $\frac{4}{5}$ of how many times $\frac{1}{4}$ of 24?
26. $\frac{3}{5}$ of 40 is $\frac{6}{8}$ of how many times $3\frac{1}{5}$?
27. $\frac{2}{5}$ is $\frac{2}{3}$ of what number?
28. $\frac{8}{9}$ is $\frac{4}{5}$ of what number?
29. $\frac{10}{11}$ is $\frac{5}{8}$ of what number?
30. $\frac{2}{3}$ is $\frac{7}{9}$ of what number?
31. $1\frac{2}{3}$ is $\frac{10}{13}$ of what number?
32. $1\frac{5}{9}$ is $\frac{7}{8}$ of what number?
33. $\frac{2}{3}$ is $\frac{4}{5}$ of how many times $\frac{5}{6}$?

MODEL. If $\frac{2}{3}$ is *four* fifths of a certain number, *one* fifth of it is $\frac{1}{4}$ of $\frac{2}{3}$, or $\frac{3}{20}$; and *three* thirds, or the whole, are 3 times $\frac{3}{20}$, or $\frac{9}{20}$. $\frac{9}{20}$ is as many times $\frac{5}{6}$ as $\frac{5}{6}$ is contained times in $\frac{9}{20}$. One sixth is contained in $\frac{9}{20}$ 6 times $\frac{9}{20}$, or $\frac{54}{20}$, which equals $\frac{27}{10}$, times; and *five* sixths is contained in it $\frac{1}{5}$ of $\frac{27}{10}$ times, or $\frac{27}{50}$. *Ans.* $\frac{27}{50}$.

34. $\frac{19}{25}$ is $\frac{4}{5}$ of how many times $1\frac{1}{5}$?
35. $\frac{7}{8}$ of $1\frac{1}{8}$ is $\frac{3}{4}$ of how many times $1\frac{1}{4}$?
36. $\frac{2}{3}$ of $\frac{9}{10}$ is 4 times what number?
37. $\frac{3}{4}$ of $1\frac{1}{11}$ is $\frac{1}{2}$ of $\frac{4}{5}$ of what number?
38. $\frac{2}{3}$ of $2\frac{4}{7}$ is $\frac{1}{3}$ of $1\frac{3}{5}$ of how many times 2?
39. $\frac{3}{8}$ of $3\frac{3}{7}$ is $\frac{9}{10}$ of how many times $\frac{4}{7}$?

40. If a grocer buys cheese for 14⅔ cents a pound, and sells it for 16 cents, how many pounds will he have to sell in order to make 7¼ cents?

41. Suppose that it takes 2⅝ yards of merino to make a sack, and 7¼ yards to make a dress. After 3 sacks have been cut from a piece containing 22¾ yards, how many dresses can be cut from what remains?

CHAPTER SIXTH.

FEDERAL MONEY.

SECTION 37.—1. One hundred cents make $1; how many cents are there in $5?

MODEL. Since there are 100 cents in $1, in $5 there are 5 times 100 cents, or 500 cents. *Ans.* 500 cents.

2. One hundred cents make $1; how many dollars in 500 cents?

MODEL. Since there are 100 cents in $1, in 500 cents there are as many dollars as 100 cents are contained times in 500 cents, or 5. *Ans.* $5.

3. When we change $5 to 500 cents, do we alter the value? Do we alter the form? What is this process called? *Ans.* **Reduction.**

4. When we change dollars to cents, do we go to a higher or lower denomination? What is this kind of Reduction called? *Ans.* **Reduction Descending.**

5. When we change cents to dollars, do we go to a higher or lower denomination? What is this kind of Reduction called? *Ans.* **Reduction Ascending.**

SECTION 38.—What is **Federal Money?** *Ans.* Federal Money is the currency of the United States.

TABLE OF FEDERAL MONEY.

10 mills (m.) make	1 cent, ...	c., ct.
10 cents,	1 dime, ...	di.
10 dimes,	1 dollar, . . .	$.
10 dollars,	1 eagle, ...	E.

Accounts are kept in dollars and cents. Cents are written at the right of dollars, with a period between, and *occupy two places.* The first place is filled with a naught, if the cents are expressed by but one figure. Mills are written at the right of cents. Thus:—

Nine dollars, 5 cents, ...	$9.05
Nine dollars, fifty cents, ...	$9.50
Nine dollars, fifty cents, five mills, . .	$9.505
Nine dollars, five mills, ...	$9.005

As 100 cents make a dollar, 1 cent is $\frac{1}{100}$ of a dollar, 2 cents $\frac{2}{100}$, &c. Cents are sometimes written as hundredths of a dollar; five dollars and twenty-five cents may be written 5\frac{25}{100}$.

1. How many mills in 3 cents?* In 7c.? In 19c.?
2. How many cents in 5 dimes? In 11 dimes?
3. How many dimes in $2? In $8? In $10?
4. How many dollars in 4 eagles? In 12 eagles?
5. How many dollars in 50 dimes?* In 170 dimes?
6. How many eagles in $60? In $90? In $20?
7. How many dimes in 50 cents? In 90c.? In 40c.?
8. How many cents in $5?

MODEL. Since 10 cents make 1 dime, and 10 dimes make $1, in $1 there must be 10 times 10, or 100, cents; and in $5, 5 times 100, or 500, cents. *Ans.* 500 cents.

---

* See MODELS, on page 72.

9. How many cents in $9? In 4 eagles? How many dimes in 7 eagles? How many mills in 3 dimes? In $2?

10. How many cents in 6 dollars 35 cents?

Model. In $1 there are 100 cents, and in $6 six times 100 cents, or 600 cents. 600 cents + 35 cents = 635 cents. *Ans.* 635c.

In such cases, to reduce to cents we need only remove the period and dollar-mark:—

$$\$6.35 = 635 \text{ cents}$$

11. How many cents in $4.89? In $10.10? In $9.05? How many mills in $2.375? In 87c. 5m.?

12. How many dollars in 635 cents?

Model. 100 cents make $1, and 635 cents will make as many dollars as 100 cents are contained times in 635 cents, or $6\frac{35}{100}$. *Ans.* $\$6\frac{35}{100}$, or $6.35.

In such cases, we need only cut off the two right-hand figures for cents, and what remains on the left will be dollars:—

$$635 \text{ cents} = \$6.35$$

13. How many dollars in 757 cents? In 843 cents? In 926 cents? In 4270 mills?

14. How many eagles in $17? How many cents in 53 mills? How many dimes in 47 cents?

15. How many cents in half a dime? In $19\frac{1}{2}$ dimes? In $5\frac{3}{4}$ dollars? How many dollars in 5 half-eagles?

16. How many cents in half a dollar? In $\frac{1}{5}$ of a dollar? In $\frac{1}{4}$ of $1? In $\frac{1}{5}$? In $\frac{1}{6}$? In $\frac{1}{8}$?

17. What part of a dollar is 50 cents?

Model. Since 100 cents make a dollar, 1 cent is $\frac{1}{100}$ of $1, and 50 cents are 50 times $\frac{1}{100}$, or $\frac{50}{100}$, which equals $\frac{1}{2}$. *Ans.* $\frac{1}{2}$ of a dollar.

18. What part of a dime is 3 cents? What part of an eagle is $4? What part of a cent is 6 mills?

19. What part of a dollar is 75 cents? $33\frac{1}{3}$ cents? 10 cents? $12\frac{1}{2}$ cents? 25 cents? 20 cents? $16\frac{2}{3}$ cents?

20. What cost 44 Grammars, at 75c. apiece?

MODEL. 75c. is $\frac{3}{4}$ of $1. 44 Grammars, at $1 apiece, would cost $44, and at $\frac{3}{4}$ of a dollar, they cost $\frac{3}{4}$ of $44, or $33. *Ans.* $33.

21. What cost 66 rulers, at $12\frac{1}{2}$ cents apiece?
22. What cost 2 dozen Readers, at $33\frac{1}{3}$c. each?
23. At 25 cents each, what cost 18 slates?
24. At $16\frac{2}{3}$ cents each, what cost 40 magazines?
25. At 50c. each, how many books can I buy for $5?
26. At $33\frac{1}{3}$c. each, how many balls can I buy for $9?

CHAPTER SEVENTH.

REDUCTION.

SECTION 39.—1. What is **Reduction**? *Ans.* Reduction is changing a quantity from one denomination to another, without altering its value.

2. What is Reduction Ascending? *Ans.* Reduction Ascending is changing a quantity from a lower denomination to a higher, without altering its value.

3. What operation do we use in Reduction Ascending? *Ans.* Division.

4. What is Reduction Descending? *Ans.* Reduction Descending is changing a quantity from a higher denomination to a lower, without altering its value.

5. What operation do we use in Reduction Descending? *Ans.* Multiplication.

SECTION 40.—What is **English** or **Sterling Money**?
Ans. The currency of Great Britain.

Table of Sterling Money.

4 farthings (far., qr.),	1 penny, . .	d.
12 pence,	1 shilling, . .	s.
20 shillings,	1 pound, . .	£.
21 shillings,	1 guinea, . .	guin.

1. How many pence in £1 5s. 6d.?

Model. In £1 are 20s. 20s.+5s.=25s. In 1s. are 12d., and in 25s. 25 times 12d., or 300d. 300d.+6d.=306d. *Ans.* 306d.

2. How many pounds, &c., in 306 pence?

Model. 12d. make 1s.; 306d. will therefore make as many shillings as 12 is contained times in 306, or 25s. and 6d. over. 20s. make £1; 25s. will therefore make as many pounds as 20 is contained times in 25, or £1 and 5s. over. *Ans.* £1 5s. 6d.

3. How many shillings in £9 3s.? In 4 guin. 7s.?
4. How many farthings in 9s. 8d.? In 11d. 3qr.?
5. How many pounds, &c., in 73s.? In 251d.?
6. How many pence in £5? In £6 11d.?
7. How many shillings, &c., in 89 far.? In 5 guin.?
8. How many pence in 1 guin. 4s.? In $\frac{1}{4}$ of £1?
9. How many farthings in 3s. $9\frac{3}{4}$d.? In $11\frac{1}{2}$d.?
10. How many pence in £1? Farthings in 1s.?
11. What part of a pound is 5s.? 4s.? 10s.?
12. What part of a shilling is 4d.? 8d.? 9d.?
13. At 6d. a pound, what cost 40 pounds of meat?

Note. Reduce the answers to the highest denomination possible.

14. What cost 8 knives, at 3s. each?
15. What cost 6 dozen slates, at 8d. each?

TROY WEIGHT.

SECTION 41.—For what is **Troy Weight** used?
Ans. For weighing gold, silver, and precious stones.

TABLE OF TROY WEIGHT.

24 grains (gr.) make 1 pennyweight, pwt.
20 pennyweights,   1 ounce, . . . oz.
12 ounces,   1 pound, . . lb.

1. Reduce 86 oz. to pounds, &c. To pennyweights.
2. How many grains in 1 oz.? In 5 pwt. 20 gr.?
3. How many pounds in 250 pwt.? In 100 oz.?
4. How many grains in 4 oz. 3 pwt.?
5. How many ounces in $\frac{1}{4}$ of a lb.? In $5\frac{5}{6}$ lb.?
6. How many pennyweights in $\frac{1}{2}$ lb.? In $4\frac{1}{4}$ oz.?
7. How many grains in $\frac{3}{4}$ of a pwt.? In $2\frac{5}{6}$ pwt.?
8. How many grains in $\frac{1}{6}$ of an ounce? In $\frac{2}{3}$ pwt.?
9. What part of a pwt. is 18 grains? 14 grains?
10. What part of an ounce is 15 pwt.? 1 grain?
11. What part of an ounce is $\frac{1}{20}$ of a pound?

MODEL. 1 lb. is 12 oz.; and $\frac{1}{20}$ of a lb. is $\frac{1}{20}$ of 12 oz., or $\frac{12}{20}$ of 1 oz. $\frac{12}{20} = \frac{3}{5}$. *Ans.* $\frac{3}{5}$ oz.

12. What part of a pennyweight is $\frac{1}{30}$ of an ounce?
13. What part of a grain is $\frac{1}{30}$ of a pennyweight?
14. What part of a pound is 120 pwt.?
15. A jeweller bought 7 pwt. of pure gold, 9 pwt. of gold coin, and 16 pwt. of silver; how many ounces did he buy altogether?
16. What is the weight in pounds, &c., of 12 spoons, of 30 pwt. each?
17. How many spoons, weighing 35 pwt. each, can be made out of 1 lb. 2 oz. of silver?

REDUCTION.

SECTION 42.—By whom is **Apothecaries' Weight** used? *Ans.* By apothecaries, in mixing medicines.

Table of Apothecaries' Weight.

20 grains (gr.) make	1 scruple,	sc. or ℈.
3 scruples,	1 dram,	dr. or ℨ.
8 drams,	1 ounce,	oz. or ℥.
12 ounces,	1 pound,	lb. or ℔.

1. How many scruples in 5 oz.? In 1 lb.?
2. How many drams in 95 gr.? In 4 lb. 5 oz.?
3. How many grains in 3 dr. 5 sc.? In $\frac{1}{4}$ lb.?
4. How many pounds in 113 ℨ? In 113 ℥?
5. Reduce 500 gr. to ounces, &c. $\frac{5}{8}$ oz. to scruples.
6. What part of a scruple is 17 gr.? 8 gr.?
7. What part of an ounce is 3 drams? 2 sc.?
8. What part of a pound is 5 oz.? $\frac{1}{2}$ dram?
9. What fraction of a dram is $\frac{1}{12}$ of an ounce?
10. What part of a grain is $\frac{1}{60}$ of a scruple?
11. How much are $\frac{3}{4}$ of a dram and $\frac{1}{3}$ of an ounce?

Model. In 1 oz. are 8 dr., and in $\frac{1}{3}$ of an ounce $\frac{1}{3}$ of 8 drams, or $2\frac{2}{3}$ drams. $2\frac{2}{3}$ drams + $\frac{3}{4}$ dr. = $3\frac{5}{12}$ drams. *Ans.* $3\frac{5}{12}$ dr.

12. How much are $\frac{3}{8}$ of a lb. and $\frac{1}{10}$ of an ounce?
13. How much are $\frac{2}{3}$ of a dram and $3\frac{1}{4}$ sc.?
14. How much are $\frac{2}{3}$ of an ounce and $\frac{1}{3}$ of a dram?
15. How many drams, &c., will it take for 20 powders, each containing 20 grains?
16. How many powders, of $1\frac{1}{2}$ sc. each, can be put up from an ounce of soda?
17. If a druggist charges 50c. for ten powders, containing 15 gr. each, at what rate is that per ounce?

AVOIRDUPOIS WEIGHT.

SECTION 43.—For what is **Avoirdupois Weight** used? *Ans.* For weighing groceries, meat, coal, cotton, drugs when sold in quantities, and all articles except gold, silver and precious stones.

TABLE OF AVOIRDUPOIS WEIGHT.

16 drams (dr.) make	1 ounce, . . .	oz.
16 ounces,	1 pound, ...	lb.
25 pounds,	1 quarter, . . .	qr.
4 quarters,	1 hundred-weight,	cwt.
20 hundred-weight,	1 ton, ...	T.

1. How many drams in a pound? Ounces in a quarter? Pounds in 1 cwt.? Pounds in a ton?

2. How many pounds in 5 T. 3 cwt. 21 lb? In $\frac{1}{4}$ of a ton? In 35 ounces? In $4\frac{1}{8}$ T.?

3. How many ounces in $5\frac{1}{4}$ lb.? In 83 dr.? In 2 cwt. 10 lb. 5 oz.? In 3 qr. 15 lb?

4. In 12 bales of cotton, averaging 400 lb. each, how many tons, &c.?

5. How many tons in 9 hogsheads of sugar, containing an average of 1000 pounds each?

6. How many four-ounce weights can be made out of $2\frac{1}{4}$ pounds of brass?

7. How many seven-pound packages of flour can a grocer put up from 2 cwt.?

8. What cost 2 cwt. of cheese, at $16\frac{1}{2}$c. a pound?

9. Bought 500 lb. of straw, at 75c. per cwt., and 400 lb. of hay, at $1.25 per cwt.; what was the bill?

10. Bought 1 cwt. of meat for $17.50; sold it at 22 cents a pound. What was the whole selling price? What was the profit?

REDUCTION.

SECTION 44.—What articles are weighed by Troy Weight? By Avoirdupois Weight? By Apothecaries' Weight? What kind of pounds (Troy, Avoirdupois, or Apothecaries') are those in the following Table?

Miscellaneous Table.

14 pounds, . .	1 stone of iron or lead.
60 pounds, . .	1 bushel of wheat.
100 pounds, . .	1 quintal of dried fish.
100 pounds, . .	1 cask of raisins.
196 pounds, . .	1 barrel of flour.
200 pounds, . .	1 bar. of beef, pork, or fish.

1. How many pounds in $5\frac{1}{2}$ stone? In $3\frac{3}{10}$ quintals of cod-fish? In a quarter of a barrel of flour?

2. Which is greater, 7 stone or 1 cwt., and how much?

3. How many bushels in 3000 lb. of wheat?

4. What cost 4 quintals of fish, at $6\frac{1}{2}$c. a pound?

5. What part of 1 cwt. is 1 stone?

6. What part of a ton is 5 cwt.? 1200 lb.?

7. What part of an avoirdupois pound is 2 ounces? What part of a Troy pound is 2 ounces? What part of an apothecaries' pound is 2 ounces?

8. What fraction of an avoirdupois ounce is $\frac{1}{20}$ of a pound? What part of a Troy ounce?

9. How do the ounce and pound of Troy Weight compare with those of Apothecaries' Weight? *Ans.* They are the same.

10. How do the ounce and pound of Troy Weight compare with those of Avoirdupois Weight? *Ans.* The Troy ounce is greater, the Troy pound less.

LONG MEASURE.

SECTION 45.—For what is **Long Measure** used? *Ans.* For measuring length or distance.

---
1 inch.

Table of Long Measure.

12 inches (in.) make 1 foot, ... ft.
 3 feet,   1 yard, ... yd.
 5½ yards,   1 rod, ... rd.
40 rods,   1 furlong, . . . fur.
 8 furlongs,   1 mile, ... mi.

1. How many inches in 3½ ft. ? In 2 yd. ?
2. How many inches in 4 yd. 1 ft. 5 in. ?
3. How many yards, &c., in 91 ft. ? In 80 inches ?
4. How many rods in a mile? How many yards?
5. How many feet in a mile? In half a mile?
6. How many miles in 480 rods? In 18 furlongs?
7. How many feet in 2 rods? In 99 inches?
8. How long will it take a person, walking at the rate of 20 rods a minute, to go 2 miles ?
9. 4 inches make a hand. What is the height in feet of a horse 15½ hands high?
10. What is used in measuring drygoods? *Ans.* The yard of long measure, divided into halves, quarters, eighths, and sixteenths.
11. Three dress-patterns, 9½, 10¼, and 8⅛ yards long, were cut from a piece containing 36 yards; how many yards were left? What did the three dresses cost, at 60 cents a yard?
12. How many half-yards of velvet can be cut from a piece 9½ yards long? How many eighths?

REDUCTION.

SECTION 46.—1. For what is **Square Measure** used? *Ans.* For measuring surfaces; such as land, walls, floors, &c.

2. What is a **Square?** *Ans.* A Square is a figure that has four equal sides perpendicular one to another—that is, leaning no more to one side than to the other.

3. What is a Square Inch? *Ans.* A square whose sides are each an inch long.

A SQUARE INCH.

1 inch.
1 inch.
1 inch.
1 inch.

TABLE OF SQUARE MEASURE.

144 square inches (sq. in.), 1 square foot, sq. ft.
  9 square feet,   1 square yard, sq. yd.
30¼ square yards,   1 square rod, sq. rd.
 40 square rods,   1 rood, . . R.
  4 roods,   1 acre, . . . A.
640 acres,   1 square mile, sq. mi.

4. How many square rods in 2 A. 3 R. ?
5. What part of a square foot is 36 sq. in. ?
6. How many square feet in 7½ sq. yd. ?
7. Reduce ½ sq. rd. to square yards.
8. How many 80-acre farms will 1 sq. mi. make?
9. Reduce 8 square rods to square yards.
10. What part of a rood is $\frac{1}{30}$ of an acre?
11. How many square rods in ¼ A. and ⅜ R. ?
12. A person, having 20 A. of land, sold it off in lots of 10 sq. rd. How many lots did it make?
13. What will it cost to plaster 288 sq. ft., at 25 cents a square yard?

CUBIC MEASURE.

SECTION 47.—1. For what is Cubic Measure used? *Ans.* For measuring bodies, which have length, breadth, and thickness; such as timber, earth, boxes, &c.

2. What is a **Cube?** *Ans.* A Cube is a body bounded by six equal squares.

3. What is a Cubic Inch? *Ans.* A cube, one inch long, one inch broad, and one inch thick. Each of its six sides is a square inch.

4. What is a Cord? *Ans.* A Cord is a pile of wood 8 ft. long, 4 ft. wide, and 4 ft. high.

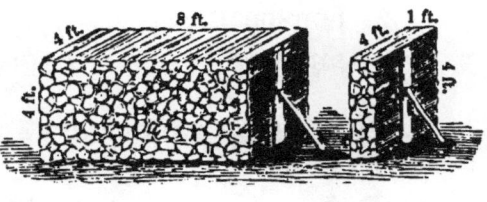

One foot in length of such a pile is called a Cord Foot.

TABLE OF CUBIC MEASURE.

1728 cubic inches (cu. in.),	1 cubic foot, cu. ft.
27 cubic feet,	1 cubic yard, cu. yd.
40 cu. ft. of round, or 50 cu. ft. of hewn timber,	1 ton or load, T.
16 cubic feet,	1 cord foot, cd. ft.
8 cord feet,	1 cord, . . Cd.

5. How many cords in a pile of wood, 24 feet long, 4 feet wide, and 4 feet high?

MODEL. 1 cord being 8 ft. long, 4 ft. wide, and 4 ft. high, there are as many cords as $8 \times 4 \times 4$ is contained times in $24 \times 4 \times 4$, or 3. *Ans.* 3 Cd.

6. How many cords in a pile of wood, 36 feet long, 4 feet wide, and 8 feet high?

REDUCTION.

SECTION 48.—For what is **Liquid** or **Wine Measure** used? *Ans.* For measuring liquids generally.

Table of Liquid Measure.

4	gills (gi.) make	1 pint, ...	pt.
2	pints,	1 quart, ...	qt.
4	quarts,	1 gallon, . . .	gal.
31½	gallons,	1 barrel, ...	bar.
2	barrels (63 gal.),	1 hogshead, . .	hhd.
2	hogsheads,	1 pipe, ...	pi.
2	pipes,	1 tun, ...	tun.

1. How many gallons in 45 pt.? In 150 gi.?
2. How many quarts in 101 gi.? In 2 bar.?
3. Reduce 5 gal. 1 pt. 1 gi. to gills.
4. Reduce $\frac{2}{3}$ of a gallon to pints. To gills.
5. Reduce 3 qt. 1 pt. to gills.
6. What part of a gallon is 1 qt.? 1 pt.?
7. Add ¼ gal., ½ qt., and ½ pt.
8. How many gallons in 5 hhd.? In 8 bar.?
9. If a tumbler holds half a pint, how many times will 1 gallon of water fill it?
10. Bought 5 gal. of oil for $10, what is the price per quart?
11. Bought a pint of milk for 4 cents; at what rate is that per gallon?
12. What cost 18 quarts of wine, at 16s. per gal.?
13. 3 pt. of molasses having been sold out of 2 gal., what is the remainder worth at 80c. a gal.?
14. If 5 bottles hold a gallon of wine, what part of a quart does each hold?

DRY MEASURE.

SECTION 49.—For what is **Dry Measure** used? *Ans.* For measuring grain, vegetables, salt, coal, and other articles not liquid.

Table of Dry Measure.

2 pints (pt.) make 1 quart, ... qt.
   8 quarts,   1 peck, ... pk.
   4 pecks,   1 bushel, . . . bu.
 36 bushels,   1 chaldron, . . . chal.

1. How many bushels in 165 qt.? In 191 pt.?
2. Reduce 1 bu. 5 qt. to quarts.
3. How many bushels in $5\frac{1}{2}$ chaldrons?
4. How many pints in 3 bu. 3 pk. 3 qt.?
5. Reduce $\frac{9}{10}$ of a peck to lower denominations.

Model. 1 pk.=8 qt. Hence, $\frac{9}{10}$ of a peck is $\frac{9}{10}$ of 8 qt., or $\frac{72}{10}$ of 1 qt., which equals $7\frac{1}{5}$ qt. 1 qt.=2 pt. Hence, $\frac{1}{5}$ of a quart is $\frac{1}{5}$ of 2 pt., or $\frac{2}{5}$ of 1 pt. *Ans.* 7 qt. $\frac{2}{5}$ pt.

6. Reduce $\frac{5}{16}$ of a bushel to lower denominations.
7. Reduce $\frac{2}{3}$ of a peck to lower denominations.
8. How many bushels in $5\frac{4}{5}$ chaldrons?
9. If a chaldron of coal costs $12, what is the price per bushel?
10. How many baskets, holding $2\frac{1}{2}$ pecks each, will 5 bushels of peaches fill?
11. What part of a bushel is half a peck?
12. What part of a quart is $\frac{1}{40}$ of a peck?
13. 6 quarts are $\frac{3}{4}$ of how many pecks?
14. If a horse is fed 6 qt. of oats a day, how long will it take him to consume 3 bushels?
15. If a family consume 4 bushels of coal a day, how long will it take them to use $3\frac{1}{2}$ chaldrons?

REDUCTION.

SECTION 50.—1. What are the natural divisions of time? *Ans.* The Year, in which the Earth revolves round the Sun; and the Day, in which it turns on its axis.

2. How is the year divided? *Ans.* Into twelve calendar months, differing in length.

3. How is the day divided? *Ans.* Into hours, minutes, and seconds.

Table of Time Measure.

60 seconds (sec.) make	1 minute, . . min.
60 minutes,	1 hour, . . . h.
24 hours,	1 day, . . . da.
7 days,	1 week, . . . wk.
365 days, or 12 calendar months,	1 year, . . . yr.
366 days,	1 leap year.
100 years,	1 century, . . cen.

4. Learn the names of the calendar months, and the number of days they contain:—

	DAYS.		DAYS.
1st month, January,	31.	7th month, July,	31.
2d month, February,	28.	8th month, August,	31.
3d month, March,	31.	9th month, September,	30.
4th month, April,	30.	10th month, October,	31.
5th month, May,	31.	11th month, November,	30.
6th month, June,	30.	12th month, December,	31.

5. How many hours in the month of April?
6. How many minutes in a day? In 2 days 3 h.?
7. How many seconds in 2 h.? In 1 h. 10 min.?
8. How many weeks, &c., in 240 hours?

SECTION 51.—Table of Paper Measure.

24 sheets make 1 quire.
20 quires, 1 ream.
2 reams, 1 bundle.
5 bundles, 1 bale.

Table of Collections of Units.

12 units make 1 dozen, doz.
12 dozen, 1 gross.
12 gross, 1 great gross.
20 units, 1 score.

1. How many sheets in a ream of paper?
2. How many sheets in $8\frac{1}{4}$ quires? In $\frac{3}{8}$ of a ream?
3. How many quires will 240 sheets make?
4. How many bundles will 100 quires make?
5. What part of a ream is 37 sheets? 15 quires?
6. What part of a gross is 5 dozen? 1 score?
7. How many units in a great gross? In 11 gross?
8. If a box of pens holds a gross, how many pens are there in 5 boxes? At $\frac{1}{4}$c. a pen, what will each box cost?
9. If a person buys a ream of paper for $3, and retails it at 1 cent a sheet, how much profit will he make?
10. If 5 dozen buttons are sold out of a gross that cost $2.40, how much are what remain worth?
11. Bought 7 quires of foolscap, 12 sheets of letter-paper, and $2\frac{1}{2}$ quires of note-paper; what was it worth, at the rate of $3.20 a ream?
12. How many units in 4 great gross?

REDUCTION.

SECTION 52.—1. Which is the shortest month?

2. How many days has February in leap-year? *Ans.* 29.

3. Which years are leap-years? *Ans.* Those that can be exactly divided by 4, except such of the even hundreds as can not be exactly divided by 400.

4. Name the leap-years between 1868 and 1901.

5. From January 1st to February 1st how many months? From Jan. 3d to Feb. 3d? From Jan. 7th to Oct. 7th?

6. From Jan. 5th to March 5th how many months? From Feb. 1st to Aug. 1st? From June 2d to Nov. 2d?

7. From April 6th to Oct. 6th how many months? From Aug. 9th to Nov. 9th?

8. What part of a year is 2 months? 6 months? 8 months? 3 months? 9 months? 5 months?

9. How many months in $\frac{5}{6}$ of a year? In $\frac{7}{12}$ of a year? In $\frac{7}{12}$ of a year? In $\frac{2}{3}$ of a year?

10. In business calculations, how many days are generally allowed to a month? *Ans.* 30 days.

11. In business calculations, what part of a month would we call 1 day? 2 days? 20 days? 15 days? 5 days? 24 days? 10 days?

12. In business calculations, what part of a year is a day considered? *Ans.* $\frac{1}{30}$ of $\frac{1}{12}$, or —— of a year.

13. What part of a year is 5 days considered?

14. What part of a year is a day really? What part is five days?

15. How many calendar months in a century?

16. Which of the months have 31 days? How many days has the 5th month? The 11th month?

CHAPTER EIGHTH.

THE METRIC SYSTEM.

SECTION 53.—1. What is the **Metric System**? *Ans.* A system of weights and measures in which it takes 10 of a lower denomination to make 1 of the next higher.

2. Where is the Metric System used? *Ans.* In France, Belgium, and other countries of Europe. Its use is also authorized by law in the United States.

3. What is the unit of length? *Ans.* The **Metre**, from which the Metric System has its name.

4. How long is a Metre? *Ans.* About $39\frac{37}{100}$ inches.

5. How are lower denominations formed? *Ans.* From the metre other denominations, $\frac{1}{10}$, $\frac{1}{100}$, and $\frac{1}{1000}$ as great, are formed with the prefixes *deci* (pronounced *des'e*), *centi*, and *milli*.

6. How are higher denominations formed? *Ans.* Denominations 10, 100, 1000, and ten thousand times as great as the metre, are formed with the prefixes *deca*, *hecto*, *kilo*, and *myria*.

Measures of Length.

10 mil′limetres make	1 cen′timetre	$= \frac{39}{100}$ inch.*
10 centimetres "	1 dec′imetre	$= 3\frac{94}{100}$ inches.
10 decimetres "	1 metre	$= 39\frac{37}{100}$ inches.
10 metres "	1 dec′ametre	$= 32$ ft. $9\frac{7}{10}$ in.
10 decametres "	1 hec′tometre	$= 328$ ft. 1 in.
10 hectometres "	1 kil′ometre	$= 3280$ ft. 10 in.
10 kilometres "	1 myr′iametre	$= 6\frac{21}{100}$ miles.

* The equivalents given in these Tables are not exact, but nearly so.

7. How many centimetres in a metre? In half a metre? In 5 metres? In 17 metres? In a decimetre?

8. How many kilometres in 4000 metres?

9. How many kilometres in 500 decametres?

10. What part of a decametre is a decimetre? A metre? 5 metres? 8 metres?

11. What part of a metre is a decimetre? A centimetre? What part of a hectometre is a decametre?

12. How many metres in 15 decimetres? In ⅛ of a decametre? In 250 centimetres? In ¼ of a hectometre?

13. Is a metre more or less than a yard? How many inches more?

14. About how many metres equal 1 rod?

15. What cost 4 metres of cloth, at $3⅛ a metre?

16. If a person sells 7 decimetres of velvet from a piece containing 3 metres, how much is what remains worth, at $10 a metre?

17. How many kilometres of fence will be required, to surround a square field 500 metres on each side?

18. At 75c. a metre, what will it cost to fence a square field, 30 metres on each side?

SECTION 54.—1. What is the unit of surface? *Ans.* The **Are** (pronounced *air*), a square whose side is 10 metres, and which equals 119⅔ square yards.

MEASURES OF SURFACE.

The cen'tiare is 1 square metre, or 1550 square inches.
100 centiares make 1 ARE = 119⅔ sq. yd.
100 ares " 1 hec'tare = $2\frac{47}{100}$ acres.

THE METRIC SYSTEM.

2. What is the unit of capacity? *Ans.* The **Litre** (pronounced *le'tur*), a cube equal to about $\frac{9}{10}$ of a quart of dry measure.

3. How are other denominations formed? *Ans.* As before, by means of the prefixes *deci*, *centi*, and *milli* for lower denominations, and *deca*, *hecto*, and *kilo* for higher ones.

MEASURES OF CAPACITY.

		Dry Measure.	Liquid Measure.
10 mil'lilitres,	1 cen'tilitre	$= \frac{3}{5}$ cu. in. $=$	$\frac{1}{3}$ fluid oz.
10 centilitres,	1 dec'ilitre	$= 6\frac{1}{10}$ cu. in. $=$	$\frac{4}{5}$ gill.
10 decilitres,	1 LI'TRE	$= \frac{9}{10}$ quart $=$	$1\frac{1}{20}$ quart.
10 litres,	1 dec'alitre	$= 1\frac{1}{7}$ peck $=$	$2\frac{5}{8}$ gallons.
10 decalitres,	1 hec'tolitre	$= 2\frac{4}{5}$ bushels $=$	$26\frac{41}{50}$ gallons.
10 hectolitres,	1 kil'olitre	$= 1\frac{3}{10}$ cu. yd. $=264\frac{17}{100}$ gallons.

The kilolitre (1 cubic metre), when used in measuring wood, is called the STERE, and equals about $\frac{11}{40}$ of a cord.

10 steres make 1 dec'astere $=$ about $2\frac{3}{4}$ cords.

4. How many ares in 550 centiares? In 5 hectares?

5. Are 2 hectares more or less than 5 acres, and how much?

6. What part of an are is 65 centiares?

7. What part of a litre is 5 centilitres? 7 decilitres? 2 millilitres?

8. How many litres in $17\frac{1}{2}$ decalitres? In $\frac{1}{25}$ of a hectolitre? In 70 decilitres?

9. What is the cost of $1\frac{1}{2}$ hectolitres of molasses, at 12 cents a litre?

10. How many litres of molasses, at 11 cents a litre, should be given in exchange for 3 decalitres and 3 litres of vinegar, at 8 cents a litre?

11. If a person bought 2 hectolitres of molasses, and sold 10 decalitres, what was the rest worth, at 12½ cents a litre?

12. If one stere equals $\frac{11}{40}$ of a cord, how many steres make a cord?

MODEL. If 1 stere equals $\frac{11}{40}$ of a cord, it will take as many steres to make a whole cord, or $\frac{40}{40}$, as $\frac{11}{40}$ is contained times in $\frac{40}{40}$; that is (rejecting the denominators), as 11 is contained times in 40, or $3\frac{7}{11}$. *Ans.* $3\frac{7}{11}$ steres.

13. If 1 litre equals $\frac{9}{10}$ of a quart, dry measure, how many litres make a quart?

14. Bought a decastere of wood, and sold 4 steres of it for $8; how much was that a stere? How much was what remained worth at the same rate?

15. At $1.75 a stere, what is the cost of 2 decasteres of wood?

16. A man bought 3 hectolitres of potatoes for $9. He sold them for 40 cents a decalitre. Did he gain or lose, and how much?

17. Reduce 1 litre 1 decilitre 1 centilitre to centilitres.

18. The kilolitre being a cubic metre, how long in inches is each side of the cube it represents?

19. If 1 decastere equals 2¾ cords, how many cords do 5 decasteres equal? 5½ decasteres?

20. How many decasteres do 19¼ cords equal?

21. If a litre equals $\frac{9}{10}$ of a quart, dry measure, how many litres are there in 45 quarts?

22. What prefixes are used with the metre, &c., to form lower denominations? What prefixes are used, to form higher denominations?

THE METRIC SYSTEM.

SECTION 55.—1. What is the unit of weight? *Ans.* The **Gram,** which equals about $15\frac{43}{100}$ grains.

WEIGHTS.

10 mil'ligrams,	1 cen'tigram	=	$\frac{15}{100}$ grain.
10 centigrams,	1 dec'igram	=	$1\frac{27}{50}$ grains.
10 decigrams,	1 GRAM	=	$15\frac{43}{100}$ grains.
10 grams,	1 dec'agram	=	$\frac{7}{20}$ oz. av.
10 decagrams,	1 hec'togram	=	$3\frac{1}{2}$ oz. av.
10 hectograms,	1 kil'ogram	=	$2\frac{1}{5}$ lb. av.
10 kilograms,	1 myr'iagram	=	22 lb. av.
10 myriagrams,	1 quintal	=	$220\frac{23}{50}$ lb. av.
10 quintals,	1 tonneau	=	$1\frac{1}{10}$ tons.

2. Which is greater, the common quintal or the quintal of the Metric System? How many pounds greater?

3. Which is greater, a ton or a tonneau?

4. If a decagram equals $\frac{7}{20}$ of an ounce avoirdupois, how many decagrams will make an ounce?

5. How many kilograms in a quintal? In a tonneau? In 1 tonneau 9 quintals? In 4500 grams?

6. How many grams in $7\frac{1}{2}$ hectograms? In 900 centigrams? In 900 decagrams? In 900 decigrams?

7. If 1 kilogram equals $2\frac{1}{5}$ lb., when butter sells at $1.10 a kilogram, how many cents a pound is it?

8. What price per kilogram is equivalent to 40 cents a pound?

9. How many powders of one gram each can be put up from $\frac{1}{2}$ of a hectogram?

10. What part of a tonneau is a kilogram? 4 myriagrams?

CHAPTER NINTH.

THE COMPOUND RULES.

SECTION 56.—1. What is a **Compound Number?** *Ans.* A number containing different denominations; as, 1 foot 2 inches.

2. What is the addition of compound numbers called? *Ans.* **Compound Addition.**

3. What is the sum of £2 17s. 6d. and £3 8s. 4d. ?

MODEL. The sum of the pence is 4d. + 6d., or 10d. The sum of the shillings is 8s. + 17s., or 25s., which equals £1 5s. The sum of the pounds is £3 + £2, which, with the £1 of the last sum, makes £6. *Ans.* £6 5s. 10d.

4. What is the sum of 15s. 9d. and 3s. 6d ?

5. What is the sum of £1 3s. 10d. and £5 9s. 4d. ?

6. A boy throws 3 pk. 6 qt. of potatoes into a barrel already containing 2 pk. 3 qt. How many bushels, &c., are then in the barrel?

7. How much jalap in two bottles, the one containing 4 oz. 3 dr. 2 sc., and the other 8 oz. 2 dr. 1 sc. ?

8. To 3 gal. 1 qt. 1 pt. of alcohol is added 1 gal. 3 qt. 1 pt. of water; how much is there of the mixture?

9. In the morning a person walks 1 mi. 1 fur. 20 rd., in 25 minutes; in the afternoon he walks 4 mi. 3 fur. 30 rd., in 1 h. 35 min. 40 sec. What distance does he walk altogether, and in what time?

10. A grocer mixes three kinds of tea; 5 lb. 11 oz. of the first kind, 6 lb. 7 oz. of the second, and 3 lb. 10 oz. of the third. How many pounds in the mixture?

11. A person, having 53 A. 1 R. 25 sq. rd. of land, bought 7 A. 20 sq. rd. more; how much had he then?

12. Bought some lace for $3.17, and some calico for $2.85; what was the amount of the bill?

MODEL. The amount of the bill was the sum of $3.17 and $2.85, or $6.02. *Ans.* $6.02.

NOTE. In the case of Federal Money, add as in simple addition; if the sum is in cents, cut off the two right-hand figures for cents, and what remains on the left will be dollars.

13. If I spend $1.10 for paper, $2.80 for books, and 25 cents for pens, how much do I spend in all?

14. Bought butter for $1.14; cheese for 27c.; lard for 32c.; sugar for $1; what did the bill amount to?

15. A person received three telegrams. The first cost 90c., the second $1.45, and the third 85c. What was the cost of all three?

16. A carpenter laid out $1.25 for nails, $2.90 for bolts, and 75c. for screws; how much was his bill?

NOTE. Sometimes the items can be combined in a particular order with advantage. Thus, in the last example, the sum of $1.25 and 75c. is $2; $2 and $2.90 make $4.90.

17. A storekeeper took in 40c. in the morning, $5.33 in the afternoon, and $4.60 in the evening; how much did he take in during the day?

18. John Ray bought of T. Kipp a dozen slates for $2.16, some envelopes for $1.72, pencils for 84c., and diaries for $4.28. How much was Ray's bill?

19. A collection having been taken up, the plate was found to contain 18 cents, 21 two-cent pieces, 10 three-cent pieces, 10 five-cent stamps, 6 10-cent stamps, and 4 25-cent stamps; how much was collected?

SECTION 57.—1. What is the subtraction of compound numbers called? *Ans.* **Compound Subtraction.**

2. A jeweller, who had 1 lb. 11 oz. 4 pwt. 3 gr. of gold, used 10 oz. 15 pwt. 20 gr.; how much had he left?

MODEL. He had left the difference between 1 lb. 11 oz. 4 pwt. 3 gr. and 10 oz. 15 pwt. 20 gr. 20 gr. can not be taken from 3 gr.; we therefore take 1 of the next higher denomination (1 pwt.), reduce it to grains (24), add it to 3 gr., and then subtract. 24 gr.+3 gr.=27 gr. 27 gr.—20 gr.=7 gr.

To balance the pennyweight thus added, we now add 1 pwt. to the 15 pwt. to be subtracted, making 16 pwt. 16 pwt. can not be taken from 4 pwt.; we therefore take 1 of the next higher denomination (1 oz.), reduce it to pennyweights (20), add it to the 4 pwt., and then subtract. 20 pwt.+4 pwt.=24 pwt. 24 pwt.—16 pwt.= 8 pwt.

To balance the ounce thus added, we now add 1 oz. to the 10 oz. to be subtracted, making 11 oz. 11 oz.—11 oz.=0 oz. 1 lb.—0= 1 lb. *Ans.* 1 lb. 8 pwt. 7 gr.

3. From a bin containing 10 bu. 3 pk. 1 qt. of oats, a person took 3 bu. 2 pk. 6 qt. How much was left?

4. If from 1 day we take 19 hours 20 min., how much time will remain?

5. A person having a bill of £5 10s. to pay, has only £3 7s. 9d. How much does he lack?

6. How much of my fence remains to be built, if there is 15 rd. 5 yd. to be built in all, and 11 rd. 2 yd. 1 ft. is completed?

7. To 10 gal. 1 qt. of whiskey a liquor-merchant adds 3 qt. of water, and then sells 7 gal. 2 qt. 1 pt. of the mixture. How much remains?

8. A hardware-merchant, having on hand 1 cwt. 10 lb. of lead, buys 12 cwt. 20 lb. more, and then sells 10 cwt. 15 lb. How much has he left?

COMPOUND SUBTRACTION. 97

9. How many years, months, and days, from November 9th, 1867, to May 1st, 1868?

Note. November is the 11th month, May the 5th. We therefore take 1867 years 11 months 9 days, which is the earlier date, from 1868 years 5 months 1 day. Allow 30 days to the month. *Ans.* 5 months 22 days.

10. How many months and days from January 15th to October 3d of the same year?

11. How many months and days from July 20th to December 29th of the same year?

12. How many months and days from July 20th to December 29th of the preceding year?

13. How many years, months, and days, from August 25th, 1865, to March 13th, 1868?

14. Some mats were bought for $18.60, and sold at a loss of $3.85; how much did they bring?

Model. If they were bought for $18.60 and sold at a loss of $3.85, they brought $18.60—$3.85, or $14.75. *Ans.* $14.75.

Note. As in the addition of Federal Money, so in subtraction, multiplication, and division; if the result is in cents, cut off the two right-hand figures for cents, and what is left will be dollars.

15. A owes B $25.20, and gives him on account $17.75. If he pays the balance with a ten-dollar bill, how much change should he receive?

16. What is the profit on goods bought for $107.50, and sold for $119.25?

17. Some goods, bought for $74.50, are sold for $69.66. Does the owner gain or lose, and how much?

18. On a bill of $48, C has paid $29.25. He now pays the balance with a twenty-dollar bill; how much change should he receive?

COMPOUND MULTIPLICATION.

SECTION 58.—1. What is the multiplication of a compound number called? *Ans.* **Compound Multiplication.**

2. How many pounds, &c., in four packages, each containing 3 lb. 5 oz. 4 dr.?

<small>MODEL. Four packages, each containing 3 lb. 5 oz. 4 dr., will contain 4 times 3 lb. 5 oz. 4 dr. 4 times 4 dr. is 16 dr., or 1 oz., 4 times 5 oz. is 20 oz., and 1 oz. (the last product) makes 21 oz., or 1 lb. 5 oz. 4 times 3 lb. is 12 lb., and 1 lb. (from the last product) makes 13 lb. *Ans.* 13 lb. 5 oz.</small>

3. What cost a dozen Histories, at 3s. 8d. each?

4. If a person gives his five children each £1 10s. 6d., how much does he give them in all?

5. What are the contents of eight pitchers, if each holds 1 qt. 1 pt. 1 gi.?

6. What is the breadth of three strawberry beds, each 1 yd. 1 ft. 10 in. broad?

7. How much land in 7 fields, each containing 1 A. 15 sq. rd.?

8. What cost 8 knives, at $1.38 each? *Ans.* $11.04.

9. What cost 9 pigs, at $5.30 apiece?

10. What cost 12 caps, at $1.33½ apiece?

11. Bought 10 albums, at $2.64 each, and 5 pocket-books, at 75c. each; what did the bill come to?

12. Bought 3 pair of gloves, at $1.05 each; 4 collars, at $3.60 each; and 11 yards of calico, at 25c. per yd. What was the amount of the bill?

13. Bought 20 chickens, at 60c. each; 6 ducks, at 75c. each; and 8 turkeys, at $1.60 each. Paid on account $16.80; how much remained due? How many fifty-cent stamps will pay the balance?

COMPOUND DIVISION.

SECTION 59.—1. What is the division of a compound number called? *Ans.* **Compound Division.**

2. If a certain pipe fills a cistern in 2 h. 21 min. 25 sec., how long will it take 4 such pipes to fill it?

MODEL. If a certain pipe fills a cistern in 2 h. 21 min. 25 sec., four such pipes will fill it in $\frac{1}{4}$ of 2 h. 21 min. 25 sec.

4 is not contained in 2 h.; we therefore reduce 2 h. to minutes (120 min.), which we add to the 21 min., getting 141 min. $\frac{1}{4}$ of 141 min.=35 min., and 1 min. remainder. We reduce this remainder, 1 min., to seconds (60 sec.), and add the result to the 25 sec., getting 85 sec. $\frac{1}{4}$ of 85 sec.=$21\frac{1}{4}$ sec. *Ans.* 35 min. $21\frac{1}{4}$ sec.

3. If 1 man can do a piece of work in 4 h. 45 min., how long will it take 6 such men to do it?

4. Dividing 20 bu. 7 pk. 3 qt. of potatoes into 10 equal heaps, how much have we in each heap?

5. It took James $\frac{5}{6}$ as long to walk a certain distance as it did Andrew. If it took Andrew 1 h. 42 min., how long did it take James?

6. A certain keg contained 5 gal. 2 qt. 1 pt. of wine mixed with water. If $\frac{1}{4}$ of the mixture was water, how much wine was there in the keg?

7. A grocer had 2 cheeses, one of which weighed $\frac{2}{3}$ as much as the other. If the larger weighed 14 lb. 4 oz., what did the smaller weigh? If the smaller weighed 14 lb. 4 oz., how much did the larger weigh?

8. If a person's weekly expenses average $40.25, how much is that per day?

9. How much will 1 portfolio cost, if they are $21.60 a dozen?

10. A person bought some goods for $25.80, and sold them for $\frac{2}{3}$ of that sum; what was his loss?

11. How many bags, holding 1 bu. 1 pk. 2 qt. each, will 11 bu. 3 pk. 2 qt. of oats fill?

MODEL. As many bags as 1 bu. 1 pk. 2 qt. is contained times in 11 bu. 3 pk. 2 qt. 1 bu. 1 pk.=5 pk. 5 pk. 2 qt.=42 qt. 11 bu. 3 pk.=47 pk. 47 pk. 2 qt.=378 qt. 42 qt. are contained in 378 qt. 9 times. *Ans.* 9 bags.

12. How many books, costing 2s. 6d. each, can be bought for £1 7s. 6d.?

13. How many pitchers, holding 1 qt. 1 pt. 3 gi. each, will 4 gall. 2 qt. 1 pt. 2 gi. fill? How many will 1 gal. 3 qt. 1 pt. fill?

14. How many dishes, at $1.10, can be bought for $5.50? For $11?

CHAPTER TENTH.

MISCELLANEOUS EXAMPLES.

SECTION 60.—1. Fifteen barrels of pork, bought for $23 a barrel, were sold for $23.62. What was the profit on the lot?

2. Five yards of cloth were bought for $32.50; being damaged, they were sold at a loss of $2.75. What did they bring per yard?

3. At what price each must I sell 12 tables, in order to gain $15, if the whole were bought for $60?

4. A merchant bought 25 yd. of silk for $52.50. He sold 15 yd. of it at $2½ a yard, and the rest for $2.60 a yard. Did he gain or lose on the whole, and how much?

5. Bought 10 stoves for $150; at what price must they be sold apiece, in order to make the cost of one stove on the whole?

6. If a person buys a ton of hay for $20, and sells 5 cwt. at 80c. a cwt., and 7 cwt. at 90c., what must he sell the rest for per cwt., so as not to lose on the whole?

7. A woman buys some oranges at 2 cents apiece. She sells half of them at the rate of 2 for 3 cents, and the rest at the rate of 3 cents apiece. She makes a dollar; how many oranges had she?

8. Three persons bought some goods for $125.90, and sold them for $150.41. What was the profit of each?

9. Eight desks were bought for $25 each, and sold for $23.75 each. How much did the owner lose?

10. A person, having bought 3 barrels of flour for $28.80, let his brother have $\frac{2}{3}$ of a barrel at cost; how much was that?

11. A farm was sold for $1800, which was $\frac{4}{7}$ of its cost. What was the loss?

12. How many bushels, &c., of apples, worth 80 cents a bushel, should be given for 3 yards of cloth, worth $3.50 a yard?

13. £1 is worth 4.86\frac{65}{100}$; at this rate, how much is one shilling worth? One penny?

14. A franc is worth 19$\frac{3}{10}$c., a shilling 22c.; how many francs are equal in value to 1 shilling?

15. Three men bought a horse for $125. After keeping him 2 months 15 days, during which time they paid $20 a month for stabling, they sold him for $130. What was each man's share of the loss?

16. If 7 oranges are equal in value to 1 pine-apple, and 3 pine-apples to 28 lemons, how many oranges are 8 lemons worth?

MODEL. If 28 lemons are worth 3 pine-apples, 1 lemon is worth $\frac{1}{28}$ of 3 pine-apples, or $\frac{3}{28}$ of 1 pine-apple; and 8 lemons are worth 8 times $\frac{3}{28}$, or $\frac{24}{28}$, or $\frac{6}{7}$, of a pine-apple. But 7 oranges equal 1 pine-apple in value; hence, if 8 lemons are worth $\frac{6}{7}$ of a pine-apple, they must be worth $\frac{6}{7}$ of 7 oranges, or 6 oranges. *Ans.* 6 oranges.

17. If 2 apples are worth as much as 1 pear, and 3 melons as much as 10 pears, how many melons should be given for 80 apples?

18. A boy, having sold 1 bu. 2 pk. of blackberries at 10c. a quart, took in part payment 1 pk. 2 qt. of timothy seed, at $2.40 a bushel. How much was still due him?

19. How many guineas are £10 10s. equal to?

SECTION 61.—1. How long will it take 15 men to do a piece of work, if 5 men can do it in 9 days?

MODEL. 15 is 3 times 5. 3 times the number of men can do the work in $\frac{1}{3}$ of the time. $\frac{1}{3}$ of 9 days is 3 days. *Ans.* 3 days.

2. If 2 pipes will empty a cistern in $1\frac{1}{4}$ hours, how many such pipes will empty it in 10 min.?

3. If 2 barrels of flour will last 8 persons $6\frac{2}{3}$ weeks, how many days will it last 4 persons?

4. If a person walks $\frac{4}{5}$ of a mile in 12 minutes, how far at that rate will he walk in an hour? How far in 1 hour 36 minutes?

5. If 8 yards of muslin cost $1.92, how much will 24 yards cost?

6. How long will it take a person working 8 hours a day to paint a house, if he can do it in 10 days, working 12 hours a day?

7. If 3 tons of hay cost $57, what cost $4\frac{1}{2}$ tons?

8. How far will a locomotive, moving at the rate of 6 miles in 15 minutes, go in an hour and a half?

9. If a barrel of flour lasts 6 adults and 3 children 20 days, how long at the same rate will it last 15 adults, rating 2 children as equal to 1 adult?

10. What is the freight on 7 tons 4 cwt., at the rate of $4.50 a ton?

11. If 4 loads of hay will serve 6 horses 3 weeks, how many weeks will 5 such loads serve 9 horses?

MODEL. If 4 loads serve 6 horses 3 weeks, 1 load will serve 6 horses $\frac{1}{4}$ of 3 weeks, or $\frac{3}{4}$ of a week,—and will serve 1 horse 6 times $\frac{3}{4}$ of a week, or $4\frac{1}{2}$ weeks. Five loads will therefore serve 1 horse 5 times $4\frac{1}{2}$, or $\frac{45}{2}$, weeks,—and will serve 9 horses $\frac{1}{9}$ of $\frac{45}{2}$ weeks, or $\frac{5}{2}$, or $2\frac{1}{2}$ weeks. *Ans.* $2\frac{1}{2}$ weeks.

12. How many acres can 6 men mow in 10 days, if 4 men can mow 30 acres in 5 days?

13. If 3 men cut 12 cords of wood in 6 days, how many days will it take 4 men to cut 9 cords?

14. If it takes 12 men 5 days to do a certain piece of work, how many men will it take to do three times as much work in 10 days?

15. If the freight on 4 tons of merchandise for 20 miles is $6, how much is it, at the same rate, on 15 cwt., for 30 miles?

16. If 6 men in 4 days can build 80 rods of stone wall, how many men will be required to extend the same wall a mile further in 12 days?

SECTION 62.—1. Three fifths of the persons at a certain meeting were ladies. If there were 96 gentlemen present, how many ladies were at the meeting?

2. If $\frac{5}{8}$ of $120 is 4 times the cost of my coat, and my vest costs $\frac{1}{3}$ as much as my coat, what is the cost of both coat and vest?

3. If from a certain number increased by 7 you subtract 3, and multiply the remainder by $\frac{1}{4}$ of 20, the product is 80; what is the number?

4. P is worth $3600; and $\frac{5}{8}$ of this sum is twice the value of Q's property. How much is Q worth?

5. $\frac{4}{5}$ of Oscar's age is 8 years less than $\frac{2}{3}$ of Lucy's, and in 5 years Lucy will be 32; how old is Oscar?

6. $\frac{1}{5}$ of the men in a regiment were killed, $\frac{1}{3}$ wounded, and $\frac{1}{4}$ captured; 300 escaped uninjured. Of how many was the regiment composed?

7. F's property consists of a house, land, and stock. His house is worth $\frac{4}{9}$ of the whole, his stock $\frac{1}{3}$, and his land is worth $400. What is his whole property worth? What is his house worth? His stock?

8. James is 18 years old, and $\frac{2}{3}$ of his age is $\frac{2}{3}$ of half his brother's age; how old is his brother?

9. A is three score years and ten. If $\frac{1}{2}$ of B's age is $\frac{1}{4}$ of C's, and $\frac{1}{6}$ of C's is $\frac{1}{10}$ of A's, how old is B?

10. If to $\frac{3}{5}$ of the number of sheep in a certain flock you add 70, you will double their number; how many sheep are in the flock?

MODEL. 70 sheep equal the difference between $\frac{3}{5}$ of the flock and twice the flock, or $\frac{10}{5}$. $\frac{10}{5} - \frac{3}{5} = \frac{7}{5}$. If 70 sheep are *seven* fifths of the flock, *one* fifth is $\frac{1}{7}$ of 70, or 10; and *five* fifths, or the whole flock, are 5 times 10, or 50. *Ans.* 50 sheep.

11. John, having lost all but $\frac{2}{3}$ of his marbles, won 48 more, and by so doing doubled his original number. How many had he at first?

12. A person, having lost $\frac{2}{3}$ of his chickens, bought 76 more, and then found that he had three times as many as at first; how many was that?

13. Grace is now $\frac{1}{4}$ the age of Blanch; were she 5 years older, she would be half Blanch's age. What is the age of each?

14. Hugh spent $\frac{2}{3}$ of his money for a melon, and the rest for cherries. If the melon cost 20c., and the cherries were 6 for a cent, how many cherries did he buy?

15. From a liberty-pole 24 feet high, $\frac{2}{3}$ of the whole, less 5 feet, was sawed off; how many feet were left standing?

SECTION 63.—1. A pole increased by $\frac{1}{5}$ of its own length would be 12 ft. long; what is its length?

MODEL. The length of the pole, being $\frac{5}{5}$ of itself, when increased by $\frac{1}{5}$ of itself, must be $\frac{6}{5}$ of itself, and this we are told equals 12 ft. If *six* fifths of the length are 12 ft., *one* fifth is $\frac{1}{6}$ of 12 ft., or 2 ft.; and *five* fifths, or the whole length, are 5 times 2 ft., or 10 ft. *Ans.* 10 ft.

2. A horse was sold at a profit of $\frac{1}{6}$ of its cost. It brought $112; what was the cost?

3. A horse was sold for $\frac{5}{6}$ of its cost, and thereby a loss of $16 was incurred; what was it sold for?

4. A horse was sold for $2\frac{1}{4}$ times its cost, and a profit of $125 was thereby realized. What did the horse cost?

5. When to a lot of iron are added two other lots, $\frac{1}{7}$ and $\frac{3}{4}$ as heavy as the first, the whole weighs 106 lb. What was the weight of the first lot?

6. Twice A's age, increased by $\frac{2}{3}$ of his age, is 56 years; how old is A?

7. After losing \$200 and giving away \$450, a person found that he had $\frac{4}{5}$ of his property left. How much was his property worth?

8. One tenth of a farmer's sheep died, and $\frac{2}{3}$ were sold. On buying as many more as then remained, he had 28; what number had he originally?

9. After selling $\frac{2}{3}$ of his coal, and then $\frac{3}{4}$ of the remainder, a coal-dealer found that he had 20 tons left. How many tons had he at first?

10. A library having taken fire, $\frac{5}{6}$ of the books were burned, $\frac{1}{4}$ of what remained stolen, and only 750 saved. How many books were in the library?

11. A farmer sent $\frac{2}{3}$ of his grain to market. $\frac{1}{7}$ of the grain sent was corn; the rest, 396 bushels, was wheat. How much grain had he left?

12. $\frac{3}{5}$ of the cost of some furniture, increased by \$50, equals the selling price. If there was a profit of \$10, what did the furniture cost?

13. If from $\frac{4}{7}$ of a certain number you subtract 6, the remainder is 50; what is the number?

14. If to $\frac{9}{10}$ of a certain number you add $4\frac{1}{2}$, the sum is 54; what is the number?

15. What number is that, $\frac{5}{8}$ of which increased by $1\frac{3}{4}$ is $27\frac{3}{8}$?

16. What number is that, $\frac{3}{4}$ of which divided by 6 is $3\frac{1}{8}$?

SECTION 64.—Fractions having a common denominator are to each other as their numerators. $\frac{4}{8}$ is to $\frac{5}{8}$ as 4 is to 5.

Fractions that have not a common denominator may be reduced to other fractions that have, and are to each other as the numerators of the latter. $\frac{2}{3}$ is to $\frac{3}{4}$ as 8 is to 9, since $\frac{2}{3}=\frac{8}{12}$ and $\frac{3}{4}=\frac{9}{12}$.

1. $\frac{5}{8}$ is to $\frac{7}{8}$ as what two numbers?
2. $\frac{3}{9}$ is to $\frac{6}{9}$ as what two numbers?
3. $\frac{5}{6}$ is to $\frac{3}{4}$ as what two numbers?
4. $\frac{1}{4}$ is to $\frac{1}{6}$ as what two numbers?
5. $\frac{1}{4}$ is to $\frac{3}{8}$ as what two numbers?
6. If an estate is divided into two shares that are to each other as 4 to 3, what parts of the estate will these shares be?

Model. To find shares that are to each other as 4 to 3, we divide the whole into $4+3$ equal parts (that is, into 7 equal parts, or *sevenths*), and take *four* of these ($\frac{4}{7}$) for the first, and five ($\frac{5}{7}$) for the second. *Ans.* $\frac{4}{7}$ and $\frac{3}{7}$.

7. If we divide a number into two parts that are to each other as 3 to 7, what fractions of the number are these parts?
8. Three partners, A, B, and C, agree that A shall have $3 of their profits to B's $2 and C's $1. What part of the profits must each receive?
9. Suppose that these partners make $600, how many dollars should each receive?
10. To divide a number into parts that are to each other as 6 to 9, what fractions of it must we take?
11. Divide 45 into two parts that are to each other as 6 to 9.

12. To divide a number into three parts that are to each other as 2, 3, and 4, what fractions of the number must we take?

13. Divide 72 into three parts that are to each other as 2, 3, and 4.

14. To divide a number into parts that are to each other as $\frac{3}{4}$ and $\frac{5}{6}$, what fractions of it must we take?

15. Divide 19 into two parts that shall be to each as $\frac{3}{4}$ and $\frac{5}{6}$.

16. Two numbers, making up 28, are to each other as 5 and 9; what are the numbers?

17. Divide 60 into three parts that shall be to each other as 2, 5, and 3.

18. A and B enter into a speculation, A contributing $200 and B $300. The profit is $140; divide it between them in proportion to the sums contributed.

19. Two boats, leaving places 100 miles apart, sail toward each other, one at the rate of 12 miles an hour, and the other 8. By the time they meet, how many miles has each gone?

20. Three partners divide their profits according to the money they put in. The first put in $1000, the second $3000, the third $4000. Their profit for a year being $2800, how much should each have?

21. A man has $500 to divide between three creditors, to whom he owes respectively $1000, $1500, and $2500. They are to be paid in proportion to their claims; how much should each receive?

22. A, B, C, and D, hire a pasture for $18. A turns in 4 cows, B 6, C 3, and D 5; how much should each pay?

SECTION 65.—1. A and B make a purchase in partnership. A contributes $300 for 2 months, and B $400 for 3 months. Their profit is $450; how should it be divided between them?

MODEL. A puts in $300 for 2 months, which is equivalent to $600 for 1 month. B puts in $400 for 3 months, which is equivalent to $1200 for 1 month. The profit should therefore be divided between them in the proportion of 600 to 1200, or 1 to 2. A should have $\frac{1}{3}$ of $450, or $150; and B, $\frac{2}{3}$ of $450, or $300. *Ans.* A, $150; B, $300.

2. Two parties received $165 for digging a drain. How should it be divided between them, if the first furnished 6 laborers for 5 days, and the second 12 laborers for 3 days?

3. A pasture is hired by two persons for $16. The first turns in 10 cows for two months, the second 2 cows for six months; how much should each pay?

4. Three persons hired a pasture for $30. The first turned in 7 cows for 5 months, the second 5 cows for 4 months, and the third 25 sheep for $4\frac{1}{2}$ months. Reckoning the pasturage of 5 sheep worth that of 2 cows, how much ought each to pay?

5. Two parties contract to do some mowing for $38.50. The first works, with his three boys, 4 days; the second furnishes three men, and works with them himself, 3 days. Reckoning 2 boys as equal to 1 man, how should the pay be divided?

6. A and B gave $20.50 for a ton of hay, to be divided between them. If A paid $8.20 of the purchase money, how many hundred-weight should he take, and how many B?

SECTION 66.—1. If an orange costs twice as much as a lemon, and together they cost 6 cents, how much does each cost?

MODEL. The cost of the lemon is once itself; the cost of the orange is twice that of the lemon. Then the cost of both (given at 6 cents) must be three times that of the lemon. If 6c. is three times the cost of the lemon, the cost of the lemon must be $\frac{1}{3}$ of 6c., or 2c.; and that of the orange must be twice 2c., or 4c. *Ans.* The lemon, 2c.; the orange, 4c.

2. Robert's age is 3 times Helen's, and together they are 36. How old is each?

3. A farmer has 4 times as many sheep as cows, and of both he has 55; how many sheep has he?

4. Charles and Jane together have 29 books. If Charles has 7 more than Jane, how many has each?

5. Two boys have $\frac{44}{10}$ of $1 between them; how many cents has each, if the first has 12 cents less than the second?

6. Ida has 4 more roses than twice as many as Dora; together they have 32; how many has each?

7. A man gave $350 for a horse, wagon, and harness. He gave 3 times as much for the wagon as for the harness, and twice as much for the horse as for the wagon; what was the cost of each?

8. A and B start with equal sums of money. A gains $45, B loses $30; and together they then have $115. How much had each at first?

MODEL. If A gains $45 and B loses $30, they must then have had $45—$30, or $15, more than they had at first. If $115 is $15 more than they had at first, they must together have had $115—$15, or $100; and each must have had $\frac{1}{2}$ of $100, or $50. *Ans.* $50.

9. A counter, show-case, and desk, cost $69. The counter cost $7 less than the show-case, and the desk $8 more than the counter; what did each cost?

10. Sold a ring, watch, and breast-pin, for $110. The watch brought 5 times as much as the ring, and the pin $5 more than the ring; what did each bring?

11. A lady bought a table, easy-chair, and sofa, for $88. For the table she gave $13 more than for the chair, and for the sofa $35 more than for the table. What did she give for each?

SECTION 67.—1. Two pipes, entering a vat, will fill it, the one in 10 minutes, the other in 15 minutes. What part of the vat will each fill in 1 minute? What part will both fill in 1 minute? How many minutes will it take both pipes to fill the vat?

2. A man can do a certain job in 5 days, and a boy in 10 days. How long will it take the man and boy, working together, to do $\frac{1}{2}$ the job? To do $\frac{1}{4}$ the job?

3. A can do a job in 10 days, and B in 12 days. After A has been working 5 days, how long would it take B to finish it? After B has been working 3 days, how long would it take A to finish it?

4. A can mow a certain field in 15 hours, B in 12 hours. After they have been mowing 2 hours, they call in C, who could do it alone in the same time as B. In how many hours will the three finish it?

5. Philip can split some wood in 8 days; with Frank's help he can do it in 5 days. In how many days could Frank do it alone?

6. F, G, and H, can dig a certain cellar in 4 days; F and H can do it in 6 days, and G and H in 8 days. How long will it take each to dig it?

MODEL. If all three can dig it in 4 days, in 1 day they can dig $\frac{1}{4}$ of it; and, if F and H can dig it in 6 days, in 1 day they can dig $\frac{1}{6}$ of it. Then G in 1 day can do the difference between $\frac{1}{4}$ and $\frac{1}{6}$, or $\frac{1}{12}$; and to do $\frac{12}{12}$, or the whole, will take G as many days as $\frac{1}{12}$ is contained times in $\frac{12}{12}$, or 12 days.

Reasoning in the same way with G and H, we find that F can do $\frac{1}{8}$ of it in 1 day, or the whole in 8 days. Now, if F and H can do $\frac{1}{6}$ of it in 1 day, and F alone $\frac{1}{8}$ of it in 1 day, H can do in 1 day the difference between $\frac{1}{6}$ and $\frac{1}{8}$, or $\frac{1}{24}$, and can therefore do the whole in 24 days. *Ans.* F 8 days, G 12 days, and H 24 days.

7. Harvey, Louis, and Oliver, can thrash some rye in $2\frac{1}{2}$ days. Without Oliver's help they can do it in 4 days; how long would it take Oliver to do it alone?

8. A vat is emptied by 3 pipes in $1\frac{1}{2}$ minutes. The 1st and 3d can empty it in $2\frac{2}{3}$ minutes, and the 2d and 3d in 3 minutes. In what time will each empty it? In what time will the 1st and 2d together empty it?

9. Three casks, A, B, and C, hold 105 gal. A and B hold 65 gal., B and C 75; how much does each hold?

10. Required the contents of three casks, if the first and second hold 42 gal., the second and third 46 gal., and the first and third 44 gal.

NOTE. As the contents of each cask are taken twice, the sum of 42, 46, and 44 gal. must be twice the contents of the three casks. After thus finding the contents of the whole, proceed as before.

11. Three pipes feed a reservoir. The first and second can fill it in 3 days, the first and third in 4 days, the second and third in $2\frac{2}{3}$ days. How long will it take each to fill it?

SECTION 68.—1. A pole 3⅓ yd. long is divided into equal parts by 5 notches. How many feet and inches are there in each division?

2. Two men travel from the same point, in the same direction, one at the rate of 3 mi. 1 fur. an hour, the other at the rate of 2 mi. 7 fur. 20 rd. an hour. How far apart will they be in half a day?

3. Two men travel from the same point, in opposite directions, one at the rate of 4 mi. 6 fur. an hour, the other at the rate of 5 mi. 5 fur. an hour. How far apart will they be in 30 minutes?

4. In New England, 6s. make a dollar. How many dollars will 40 yd. of calico cost, at 9d. a yard?

MODEL. 6s. (which in N. E. is $1) equals 72d.; 9d. is therefore $\frac{9}{72}$, or $\frac{1}{8}$, of $1. At $1 a yard, 40 yd. would cost $40; and at $\frac{1}{8}$ of a dollar a yard they cost $\frac{1}{8}$ of $40, or $5. *Ans.* $5.

5. At 9s. a day, how many dollars will it cost to hire three men for four days in Massachusetts?

6. How many primers, at 9d. apiece, can be bought for $3.50 in New Hampshire?

7. If Emma can pick a quart of berries in $\frac{1}{4}$ of an hour, and Rose can pick a quart in $\frac{3}{10}$ of an hour, how many quarts can both pick in 2 h. 30 min.?

8. Two pipes can fill a vat in $\frac{1}{4}$ of an hour. The first can fill it in half an hour; how many such vats could the second fill in one hour?

9. A girl, buying some paper, needed 10c. more to get the best, which was 2c. a sheet, but had just enough to buy what was sold at the rate of 2 sheets for 3c. How many sheets did she want?

10. Edward bought 4 tops, and had 10 cents left.

Had he bought 7 such tops, he would have had but 1c. left. How much were the tops apiece?

11. A lady gave each of her children $3, and had $9 left in her purse. Had she wanted to give each $5, she would have needed $5 more. How many children had she?

12. Horatio is 14 years older than Valorus, and together they are 41. How old is each?

13. The difference of two numbers is 10, their sum is 28; what are the numbers?

14. What two numbers are those, whose sum is 125, and their difference 25?

SECTION 69.—1. A hound, running after a fox, gains 20 rods on him every minute. If the fox has half a mile the start, how long before he will be caught?

2. A deer is $\frac{1}{4}$ of a mile before a hound, and every minute the deer runs 60 rods, and the hound 70. How long will it be before the deer is overtaken?

3. A policeman runs 5 rods to a thief's 4. The thief has 10 rods the start; how far will the policeman run before he catches the thief?

MODEL. If the policeman runs 5 rods to the thief's 4, he gains 1 rod on every 5 he runs; and, to gain 10 rods, he will have to run as many times 5 rods as 1 rod is contained times in 10 rods, or 10. 10 times 5 is 50. *Ans.* 50 rods.

4. How far will the thief mentioned in the last Example run before he is caught?

MISCELLANEOUS EXAMPLES.

5. A hare is 25 leaps in advance of a hound, and takes 4 leaps to the hound's 3, but 1 leap of the hound equals 1¼ leaps of the hare. How many leaps will each take before the hare is caught?

6. A train of cars is 10 miles behind a stage. If the cars go 5 times as fast as the stage, how many miles, &c., will the stage go before it is overtaken?

7. At what time between 2 and 3 o'clock are the minute and hour hand of a watch together?

MODEL. The minute and hour hand are together at 12 o'clock. In the course of 12 hours, the minute hand will overtake the hour hand 11 times; to overtake it once, therefore, will require $\frac{1}{11}$ of 12 hours, or $1\frac{1}{11}$ hours.

When the hands are together between 2 and 3, the minute hand will have overtaken the hour hand twice since 12 o'clock, which will require twice $1\frac{1}{11}$ hours, or $2\frac{2}{11}$ hours. *Ans.* $2\frac{2}{11}$ hours past 12, or 10 min. $54\frac{6}{11}$ sec. past 2.

8. At what time between 3 and 4 will the hour and minute hand of a clock be together?

9. At what time between 7 and 8 will the hands be together? At what time between 9 and 10?

10. A person received from his employer $74, under an agreement that he was to board with the latter 40 days, and receive $2.50 a day when he worked, and lose 75c. a day when he was idle. How many days was he idle?

MODEL. His pay being $2.50 a day, if he had worked the whole 40 days, he would have received 40 times $2.50, or $100; hence he lost by being idle the difference between $100 and $74, or $26. Each day he was idle he failed to make $2.50, and forfeited 75c. besides, thus losing in all $3¼. Hence, if he lost $26, he must have been idle as many days as $3¼ is contained times in $26, or 8. *Ans.* 8 days.

SECTION 70.—1. A contracted to work 36 days for B, on condition that he should receive $2 for every day he worked, and forfeit 50c. each day he was idle. If he received from B $47, how many days was he idle?

2. How many square yards in an oblong surface 7 yards long and 4 yards wide?

MODEL. A surface 1 yd. long and 1 yd. wide contains 1 sq. yd.; a surface 7 yd. long and 1 yd. wide, would therefore contain 7 times 1, or 7, sq. yd.; and a surface 7 yd. long and 4 yd. wide contains 4 times 7, or 28, sq. yd. *Ans.* 28 sq. yd.

3. How many square rods in an oblong field 20 rods by 15 rods?

4. How many sq. yd. in a floor 12 ft. square?

5. At 30c. a square yard, what will it cost to plaster a wall 10 ft. high and 18 ft. long?

6. What will it cost to paint a wall 9 yd. long and 4 yd. high, at 5c. per square foot?

7. How many marble blocks 12 in. square will be needed to pave a walk 3 ft. wide and 20 ft. long?

8. How much yard-wide carpeting will be required to cover a room 18 ft. by 15 ft.?

9. How much carpeting $\frac{3}{4}$ of a yard wide is required for a room 9 yd. long by 14 ft. wide?

10. How many cubic feet in a box 4 ft. long, 3 ft. wide, and 2 ft. high?

MODEL. A box 1 ft. long, 1 ft. wide, and 1 ft. high, would contain 1 cu. ft.; a box, therefore, 4 ft. long, 1 ft. wide, and 1 ft. high, would contain 4 times 1, or 4, cu. ft.; a box 4 ft. long, 3 ft. wide, and 1 ft. high, would contain 3 times 4, or 12, cu. ft.; and a box 4 ft. long, 3 ft. wide, and 2 ft. high, contains twice 12, or 24, cu. ft. *Ans.* 24 cu. ft.

11. How many cubic yards in a bin 5 yd. long, 2 yd. wide, and 3 yd. high?

12. What is the freight on a box 3 ft. long, 3 ft. wide, and 3 ft. high, at 50c. a cubic foot?

13. What will it cost to dig a cellar 12 feet long, wide, and high, at 60c. a cubic yard?

CHAPTER ELEVENTH.

PERCENTAGE.

SECTION 71.—1. What do the words **Per cent.** mean? *Ans. By* or *on the hundred.*

2. What does **1 Per cent.** mean, and how may it be written? *Ans.* 1 Per cent. means *one on every hundred,* or 1 *hundredth,* and may be written 1%.

3. What does **2 Per cent.** mean, and how may it be written? *Ans.* 2 Per cent. means *two hundredths* ($\frac{2}{100}$), and may be written 2%.

4. What part is 3%? 7%? 11%? 13%?

5. What part is 50%? *Ans.* $\frac{50}{100}$, or $\frac{1}{2}$.

6. What part is 25%? What part is 20%?

7. What part is 10%? What part is 5%?

8. What part is 4%? What part is 2%?

9. What part is 6%? What part is 40%?

10. What part is 80%? What part is 35%?

11. What per cent. is $\frac{1}{2}$ equal to?

MODEL. A whole is 100% of itself, and $\frac{1}{2}$ is $\frac{1}{2}$ of 100%, or 50%. *Ans.* 50%.

12. What per cent. is $\frac{1}{8}$ equal to? $\frac{1}{4}$? $\frac{1}{12}$?
13. What per cent. is $\frac{1}{5}$ equal to? $\frac{1}{6}$? $\frac{1}{10}$?
14. What per cent. is $\frac{1}{8}$ equal to? $\frac{1}{9}$? $\frac{1}{20}$?
15. What per cent. is $\frac{2}{3}$ equal to? $\frac{4}{5}$? $\frac{1}{30}$?
16. What per cent. is $\frac{3}{4}$ equal to? $\frac{3}{4}$? $\frac{1}{25}$?
17. What part is $33\frac{1}{3}\%$? What part is $66\frac{2}{3}\%$?
18. What part is $16\frac{2}{3}\%$? What part is $12\frac{1}{2}\%$?

SECTION 72.—1. If a person loses $\frac{2}{3}$ of his property, what per cent. of it has he left?

2. If a person sells 30% of his sheep, what part of his flock does he retain?

3. How much is 5% of $500?

MODEL. $5\% = \frac{5}{100}$. $\frac{1}{100}$ of $500 is $5, and $\frac{5}{100}$ is 5 times $5, or $25. *Ans.* $25.

4. How much is 4% of $25?

MODEL. $4\% = \frac{4}{100}$, or $\frac{1}{25}$. $\frac{1}{25}$ of $25 is $1. *Ans.* $1.

NOTE. It is sometimes best to reduce the fraction to its lowest terms, as in this Example,—and sometimes not, as in the last. The pupil must exercise his judgment.

5. How much is 3% of $175? 10% of $345?
6. How much is 6% of $150? 25% of $960?
7. How much is 9% of $400? 75% of $3.60?
8. How much is 4% of $2.50? 60% of $4.75?
9. How much is 8% of $9.00? 50% of $8.88?
10. How much is $33\frac{1}{3}\%$ of £36? $12\frac{1}{2}\%$ of 8 bu.?
11. How much is 20% of 1 mi.? $16\frac{2}{3}\%$ of £4?
12. How much is $8\frac{1}{3}\%$ of 27 A.? 40% of $75?
13. How much is 7% of $1000? 11% of 25 gal.?

14. How much is 2 % of $20? 16 % of 12½ ft.?

15. The profit on certain goods, bought for $120, was 33⅓ %; how many dollars was it?

16. What must goods bought for $120 be sold for, in order to make 33⅓ %?

17. What was the profit on goods bought for $150, and sold at an advance of 20 %?

18. What was the loss on a farm bought for $1200, and sold at 5 % below cost? What did it bring?

19. A merchant, having bought 1 dozen pair of boots for $72, wishes to make 16⅔ % on them. How much must he charge a pair?

20. A and B had $200 each. A gained 6 % on his capital; B lost 12½ % on his. How much more was A then worth than B?

21. A farmer, having 80 chickens, lost 25 % of them, and then sold 75 % of the remainder; how many chickens had he left?

22. A collector, who charges 5 % commission, collects two bills, one of $30, the other of $150; how much should he retain, and how much pay over?

23. Bought some goods for $600; sold 20 % of them at a loss of 10 %, and the rest at an advance of 5 %. Did I gain or lose on the whole, and how much?

24. A hogshead of molasses, containing 60 gal., was bought for $30. 50 % of the contents having leaked out, what must the rest be sold for a gallon, so as not to lose on the whole?

25. How much is 100 % of 999? What per cent. is anything of itself?

26. How much is 200 % of 47? 300 % of 56?

SECTION 73.—1. What % of $500 is $10?

MODEL. $10 is $\frac{10}{500}$, or $\frac{1}{50}$, of $500. $\frac{1}{50} = 2\%$. Ans. 2%.

2. What % is $5 of $50? £6 of £150?
3. What % is $9 of $36? $3.30 of $16.50?
4. 1 cwt. of 1 ton (20 cwt.)? 2 dimes of $1?
5. 1 foot of 1 yard? 3 pecks of a bushel?
6. A man bought some goods for $100, and sold them at a loss of $7; what was the rate % of loss?
7. What was the rate of loss on goods bought for $100, and sold for $93?

NOTE. The rate of gain or loss must always be reckoned on the *cost*.

8. Some muslin, bought for $96, was sold for $102; what was the rate of profit? What would it have been, had the muslin brought $104?
9. A bookseller lets a teacher have a dollar book for 75c.; what % does he take off?
10. A house was put up for $3000, on a lot that cost $600. The whole was sold for $4500. Did the owner gain or lose, and what rate %?
11. An agent gets $9 for collecting a bill of $300; what % does he receive?
12. If a house rents this year for $450, which last year brought $25 a month, how much per cent. has the rent advanced?
13. 80 is 33⅓ % of what number?

MODEL. $33\frac{1}{3}\% = \frac{1}{3}$. If 80 is ⅓ of the required number, 3/3, or the whole, must be 3 times 80, or 240. *Ans.* 240.

14. 16 is 40 % of what number?
15. 15 lb. is 3 % of how many cwt.?

16. 50% of $50 is 25% of what?

17. A collector, who charges 10%, receives $40 for collecting a bill; what was its amount?

18. Selling some goods at 7% advance, a person makes $14; what did the goods cost?

19. A planter lost $21 on a horse; if his loss was at the rate of 14%, what was the cost of the horse? What did he sell the horse for?

20. Selling a house for 6% less than it cost, a person lost $120; what was the selling price?

21. 20% of £10 is 6% of what?

SECTION 74.—1. After gaining 50% of his capital, a trader had $1800; what was his capital?

Model. His capital was 100% of itself. After gaining 50% of his capital more, he had 150%, or $\frac{3}{2}$, of his capital. If $1800 is $\frac{3}{2}$ of his capital, $\frac{1}{2}$ of his capital is $\frac{1}{3}$ of $1800, or $600; and $\frac{2}{2}$, or the whole capital, is twice $600, or $1200. *Ans.* $1200.

2. A person gave $160 for an ox, which was $33\frac{1}{3}$% more than its real value; what was its value?

3. If a lot, sold for $495, brings 10% less than its value, at what price would it have brought 10% more than its value?

4. A certain number, diminished by $12\frac{1}{2}$% of itself, is 35; what is the number?

5. A man gave his daughter 25% of the rent of a certain house, and the house rented for 10% of its value. If the daughter received $50, how much was the house worth?

6. P sold a horse to Q at a profit of 5%; Q sold him to R at a profit of 10%. If R paid $231 for the horse, what did P pay?

7. A person sent his agent $840, to pay for an investment and the agent's charge of 5% on the same. What was the amount of the investment?

8. An agent, having collected some money, retained his commission, which was at the rate of 2%, and paid over $147. What was the amount collected, and how much was his commission?

9. A collected for B 20 bills of equal amount, and, after deducting 10% for his commission, paid over the balance, $108. How much was each bill?

10. In a grove of pine, spruce, and cedar trees, 25% of the trees are pines, and 35% spruces. If there are 32 cedars, how many trees does the grove contain?

11. Fifteen per cent. of my peach trees having died, I set out 39 more, and then had 50% more than at first. How many had I at first?

12. Having gained 20% on his capital, a merchant lost $160, and found that he then had in all $1400. What was his original capital?

13. A person sold 6 watches for $60 apiece, which was 25% less than they cost? How much did he lose on the whole?

14. To-day I take in $20, which is 20% less than I took yesterday, and $33\frac{1}{3}$% more than I took the day before. What are my receipts for all three days?

15. Having doubled his money, B gave away 7% of what he then had, and found that he had $279 left. How much had he at first?

BANKRUPTCY.

SECTION 75.—1. What is a **Bankrupt?** *Ans.* One who fails in business and can not pay his debts.

2. What is meant by a bankrupt's **Assets?** *Ans.* The property in his hands.

3. What is meant by a bankrupt's **Liabilities?** *Ans.* His debts or obligations.

4. If a bankrupt can pay 25 cents on the dollar, what per cent. of his debts can he pay? How much should A receive, whom he owes $200?

5. A fails, owing $9000, and having $1500 assets. What % will his creditors get on their claims?

MODEL. As A's debts are $9000, and he has $1500 to pay them with, each dollar of debt must draw $\frac{1}{9000}$ of $1500, or $\frac{1500}{9000}$ of $1. $\frac{1500}{9000} = \frac{1}{6}$, which is $16\frac{2}{3}\%$. *Ans.* $16\frac{2}{3}\%$.

6. If the bankrupt mentioned in the last Example owes C $420, how much of the assets should C receive?

7. A person fails, having four creditors; he owes A $1200, B $1500, C $300, and D $1000. His assets are $500. How many cents can he pay on the dollar, and how much should each creditor receive?

8. A bankrupt pays 75c. on the dollar; how much will a creditor lose, whom he owes $920? How much will a creditor receive, whom he owes $840?

9. If a creditor receives $420 for a debt owed him by a bankrupt who can pay but 40c. on the dollar, how much was the debt?

10. A person fails. His liabilities are $8000 owed to B, and 20% of that amount owed to C. His assets are $400 cash, $1600 in goods, and $1200 in notes. What per cent. can he pay, and how much should each of his creditors receive?

INSURANCE.

SECTION 76.—1. What is **Insurance?** *Ans.* Insurance is a contract, by which, for a certain sum paid, one party secures another against loss by fire, the dangers of navigation, &c.

2. What is the sum paid called? *Ans.* The **Premium**.

3. A paint-store is insured to the amount of $5000, at $1\frac{1}{2}$%. What is the premium?

MODEL. At 1%, the premium would be $\frac{1}{100}$ of $5000, or $50; at $\frac{1}{2}$%, it would be $\frac{1}{2}$ of $50, or $25. $50+$25=$75. *Ans.* $75.

4. A house is insured for $2000, and the furniture in it for $1000. What does the insurance cost, at $\frac{1}{4}$ of 1 per cent.?

5. Two hundred barrels of flour, worth $8 a barrel, are insured for 75% of their value. What is the premium, if the rate is 40c. on $100?

6. If a hotel is insured for $9000, at 2%, what is the premium?

7. The premium on a factory, insured for $7500, is $75; what is the rate?

8. If a person pays $200 for insuring a boat, at $2\frac{1}{2}$%, for how much does he insure her?

9. A house worth $4000 is insured for $\frac{3}{4}$ of its value, at $\frac{3}{10}$ of 1%. If it burns down, how much will the owner receive from the insurance company, and how much will he save by having insured?

10. A store and its contents are insured for 15% of their value, at 1%. If the premium is $9, and the value of the store is 200% of the value of the contents, how much is the store worth? What is the value of the contents?

TAXES.

SECTION 77.—1. What is a **Tax**? *Ans.* A Tax is a sum assessed on the person, property, or income of an individual, for the support of government.

2. What is a Poll-tax? *Ans.* A tax on the person, generally a uniform sum on each male citizen except such as are exempted by law.

3. How is a Property-tax reckoned? *Ans.* At a certain % of the estimated value of the property.

4. What is Real Estate? *Ans.* Fixed property, such as lands and houses.

5. What is Personal Property? *Ans.* That which is movable; cash, notes, furniture, &c.

6. If a tax of $90 is to be raised, and there is taxable property valued at $6000, what is the rate, and what must A pay, who has property valued at $1500?

MODEL. If $6000 worth of property has to pay a tax of $90, $1 must pay $\frac{1}{6000}$ of $90, or $\frac{90}{6000}$ of $1, or $\frac{3}{200}$. $\frac{3}{200} = 1\frac{1}{2}$ hundredths, or $1\frac{1}{2}$ per cent.

If A is taxed on $1500 worth of property, he must pay $1\frac{1}{2}$ % of $1500, or $22.50. *Ans.* $22.50.

7. The taxable property of a district is estimated at $8000; a tax of $160 is to be raised. What must A pay on $425 personal property? What must B pay on $500 personal property and $600 real estate?

8. A poll-tax of 75c. being laid, how much must a person pay who is taxed for 4 polls?

9. The tax-rate in a certain town being $\frac{1}{2}$ of 1 %, what will be the tax-bill of a person who has real estate valued at $1000 and $600 worth of personal property? What will be the bill of one who has $2200 worth of personal property?

SECTION 78.—1. What are **Duties?** *Ans.* Duties are taxes on imported goods, levied for the support of government.

2. How are duties charged? *Ans.* Either at a certain sum on each yard, gallon, &c., of the article imported; or at a certain per cent. on the cost of the article in the country where it was bought.

3. What is the duty on $4500 worth of silks, the rate being 60 %?

4. Required the duty on 5 hogsheads of molasses, containing, after leakage is deducted, 60 gallons each, the rate being 8c. a gallon.

5. What is the duty on 10 hhd. of sugar, weighing 1200 lb. each, $12\tfrac{1}{2}$ % being allowed for the weight of the casks, and the duty being 2c. a pound?

6. The duty on sewing-silk is 40 % on the cost. What is the duty on 100 lb. of sewing-silk bought for $11 a lb., and 50 lb. bought for $12 a lb.?

7. How much must sewing-silk be sold for, to gain 50 % on the entire cost, if it was bought for $12 a lb., and the duty on it was 40 %?

8. A case of brandy, containing 12 bottles, was bought for $24; the duty was $3.60 a gallon. Five bottles being allowed to the gallon, what must be charged per bottle, to make 10 % on the whole cost?

9. How many pounds of raisins does a merchant import, if the duty on them, at $2\tfrac{1}{2}$ cents a pound, amounts to $115?

10. The duty on 8 packages of snuff, averaging 10 lb. net weight each, is $40. What is the rate of duty per pound?

CHAPTER TWELFTH.

INTEREST.

SECTION 79.—1. What is **Interest?** *Ans.* Interest is what is paid for the use of money.

2. What is the **Principal?** *Ans.* The money used, for which interest is paid.

3. What is the **Rate?** *Ans.* The per cent. paid for the use of the principal for a certain time,—one year, unless some other time is specified.

4. What is the **Amount?** *Ans.* The sum of the principal and interest.

A person borrows $100 for a year, and pays $6 for its use; the Principal is $100, the Interest $6, the Rate 6 %, the Amount $106.

5. At 7 %, what is the interest of $49, for 1 yr. ?

MODEL. $7\% = \frac{7}{100}$. $\frac{1}{100}$ of $49 is 49c., and $\frac{7}{100}$ is 7 times 49c., or 343c.—which equals $3.43. *Ans.* $3.43.

6. Find the interest of $900, at 3 %, for 1 yr.
7. Find the interest of $800, at 6 %, for 1 yr.
8. Find the interest of $350, at 9 %, for 1 yr.
9. Find the interest of $700, at 7 %, for 1 yr.
10. Find the interest of $2000, at 5 %, for 1 yr.
11. Find the interest of $90.50, at 4 %, for 1 yr.
12. Find the interest of $50.50, at 8 %, for 1 yr.
13. Find the amount of $600, for 1 yr., at $5\frac{1}{2}$ %.

MODEL. At 1 %, the interest on $600 would be $6,—at 5 %, 5 times $6, or $30,—and at $\frac{1}{2}$ %, $\frac{1}{2}$ of $6, or $3. $30 + $3 = $33. The *interest*, at $5\frac{1}{2}$ %, is $33; and the *amount* is $600 + $33, or $633. *Ans.* $633.

14. Find the amount of $100, for 1 yr., at 4¾%.
15. Find the amount of $480, for 1 yr., at 6½%.
16. Find the amount of $820, for 1 yr., at 7½%.
17. Find the amount of $80, for 1 yr., at 6%.
18. Find the interest of $450, for 1 yr., at 7%.
19. Find the interest of $240, for 1 yr., at 5½%.
20. What must I pay, to take up a note for $100, with interest at 5% for 12 months?
21. A person lends $1000 for 1 year, at 7%. With another $1000 he buys a house, which at the end of a year he sells for $1100. From which investment does he make the more money, and how much more?

SECTION 80.—1. What is the interest of $200, at 7%, for 2 years?

MODEL. The interest on $200, at 7%, for 1 year, is $14; and for 2 years it is twice $14, or $28. *Ans.* $28.

2. Find the interest of $120, for 3 yr., at 5%.
3. Find the interest of $400, for 6 yr., at 7%.
4. Find the amount of $700, for 3 yr., at 6%.
5. Find the amount of $40, for 5 yr., at 4½%.
6. Find the interest of $600, for 4 yr., at 8%.
7. Find the amount of $350, for 2 yr., at 7%.
8. What is the interest of $8000, for 10 yr., at 10%? Of $300, for 20 yr., at 5%? What is the interest on any principal, when the product of the rate and years is 100?
9. Find the interest of $80, for 3 mo., at 7%.

MODEL. The interest of $80, at 7%, for 1 yr., is $5.60; and for 3 mo., which is ¼ of a year, it is ¼ of $5.60, or $1.40. *Ans.* $1.40.

INTEREST. 129

10. At 8 %, what is the interest of $1000, for 2 months? For 5 months?

11. At 7 %, what is the interest of $1600, for 3 months? For 9 months? For 1 year 9 months?

12. At 5 %, what is the amount of $300, for 2 months? For 1 yr. 2 mo.? For 2 yr. 2 mo.?

13. At 4½ %, what is the amount of $200, for 10 months? For 6 mo.? For 3 yr. 6 mo.?

14. At 7 %, what is the interest of $60, for 4 mo.? For 1 yr. 4 mo.? For 11 mo.?

15. At 4 %, what is the interest of $1000, for 1 mo.? For 5 mo.? For 8 mo.?

16. At 5½ %, what is the interest of $1200, for 7 mo.? For 1 yr. 11 mo.?

17. At 7 %, what is the interest of $2400, from October 3d to December 3d of the same year?

18. What is the interest of $750, from February 15th to August 15th of the same year, at 6 %?

19. What is the amount of $600, from November 1st to June 1st of the next year, at 5 %?

20. What must I pay for the use of $175, from May 8th, 1866, to January 8th, 1868, at 8 %?

21. February 1st A lent B $100, and on March 1st $200 more. Interest being allowed at 6 %, how much should B pay A on settling up, September 1st of the same year?

22. A person depositing $200 in a savings-bank, June 1st, 1867, gets 5 % interest. Had he waited till July 1st, he might have invested it differently at 7 %; in this case, how much better off would he have been on the 1st of June, 1868?

SECTION 81.—1. What is the interest of $600, for 10 days, at 5%?

NOTE. In calculating interest, 30 days are allowed to the month.

MODEL. At 5%, the interest of $600, for 1 year, is $30,—for 1 month, $\frac{1}{12}$ of $30, or $2.50,—and for 10 days, which is $\frac{1}{3}$ of 1 month, $\frac{1}{3}$ of $2.50, or 83$\frac{1}{3}$c. *Ans.* 83$\frac{1}{3}$c.

We reject fractions of a cent less than $\frac{1}{2}$, and call $\frac{1}{2}$ or more an additional cent; which makes the above answer 83c.

2. What is the interest of $950, at 6%, for 3 days? For 10 days?

3. What is the amount of $800, at 7%, for 12 days? For 15 days?

4. What is the amount of $2000, at 4$\frac{1}{2}$%, for 8 days? For 20 days?

5. What is the interest of $1200, at 4%, for 5 days? At 7%, for 25 days?

6. What is the interest of $900, at 5%, for 4 days? At 6%, for 18 days?

7. What is the amount of $450, at 8%, for 27 days? At 7%, for 7 days?

8. What is the interest of $7500, at 5%, for 22 days? For 28 days?

9. What is the interest of $85.50, at 8%, for 15 days? For 18 days?

10. What is the interest of $100, at 7%, from May 1st, 1868, to May 22d, 1868?

11. What is the amount of $250, at 6$\frac{1}{2}$%, from Jan. 1st, 1868, to Jan. 31st, 1868?

12. Borrowed, May 28th, $200, and June 3d, $300. How much money will be needed to pay off these debts, with interest at 7%, June 12th of the same year?

INTEREST.

SECTION 82.—1. What is the interest of $300, at 7 %, for 1 yr. 6 mo. 6 da. ?

MODEL. At 7 %, the interest of $300, for 1 year, is $21. For 1 month, the interest is $\frac{1}{12}$ of $21, or $1.75,—for 6 months, 6 times $1.75, or $10.50,—and for 6 days, which is $\frac{1}{5}$ of a month, $\frac{1}{5}$ of $1.75, or 35c. For 1 year 6 months 6 days, therefore, the interest is $21 + $10.50 + 35c., or $31.85. *Ans.* $31.85.

2. What is the interest of $900, at 7 %, for 1 year 4 months 10 days?

3. What is the amount of $480, at 6 %, for 9 mo. 5 da.? At 8 %, for the same time?

4. At 4 %, what is the amount of $1020 for 1 yr. 8 mo. 9 da.? For 2 yr. 8 mo. 9 da.?

5. At $5\frac{1}{2}$ %, what is the interest of $1000, for 1 year 3 days? For 3 yr. 3 mo.?

6. At 5 %, what is the interest of $3200 for 2 yr. 3 mo. 18 da.? For 3 yr. 9 mo. 1 day?

7. At 1 % a month, what is the interest of $5000, for 21 days? For 5 mo. 2 da.? For 2 mo. 27 da.?

8. At $\frac{3}{4}$ of 1 % a month, what is the interest of $800 for 3 mo. 28 da.? For 1 mo. 15 da.?

9. What is the amount of $625, at 10 %, from October 1st, 1866, to April 4th, 1868?

10. A person bought some property, Jan. 1st, 1868, for $1000, borrowing the money to pay for it, at 6 %. For how much must he sell it July 1st, 1868, to make 10 % on the cost and interest?

11. A house was bought for $8000,—$2000 cash, and the balance to be paid in two equal semi-annual instalments, with interest at 7 %. What was the whole amount paid?

INTEREST.

SECTION 83.—1. What is meant by the **Legal Rate** of interest? *Ans.* A rate fixed by law, for cases in which no rate is specified.

2. In what States is the legal rate 7 %? *Ans.* In New York, New Jersey, Michigan, Wisconsin, Minnesota, Kansas, South Carolina, and Georgia.

3. What is the legal rate in most of the United States? *Ans.* In most of the United States, including all the New England States, the legal rate is 6 %. Hence it is important to know the shortest methods of computing interest at 6 %.

4. At 6 %, what is the interest of $1, for 2 months? At 6 %, the interest for 2 months is what part of the principal?

5. Find the interest of $75, at 6 %, for 2 mo.

MODEL. At 6 %, the interest for 2 months is $\frac{1}{100}$ of the principal. $\frac{1}{100}$ of $75 is 75c. *Ans.* 75c.

6. Find the interest of $8300, at 6 %, for 2 mo.
7. Find the interest of $302.50, for 2 mo., at 6 %.

NOTE. Dividing a sum consisting of dollars and cents by 100 is done by simply moving the point two places to the left. $\frac{1}{100}$ of $302.50 is $3.025. *Ans.* $3.03.

8. Find the amount of $60.25, for 2 mo., at 6 %.
9. Find the interest of $999, for 2 mo., at 6 %.
10. At 6 %, what is the interest of $130.75, for 2 months? The amount of $8497, for 60 days?
11. Find the interest of $2450, for 16 mo., at 6 %.

MODEL. The interest for 2 months is $\frac{1}{100}$ of the principal, or $24.50; and for 16 months it is as many times $24.50 as 2 mo. is contained times in 16 mo., or 8. 8 times $24.50 is $196. *Ans.* $196.

12. At 6 %, what is the interest of $88.90, for 8 months? For 18 months? For 1 year 6 months?

13. At 6 %, what is the amount of $9000, for 10 months? For 1 year 2 months?

14. At 6 %, what is the interest of $4800, for 11 months? For 7 months? For 2 yr. 8 mo.?

15. At 6 %, what is the amount of $145, for 20 mo.?

16. At 6 %, what was the interest of $700, from July 7th, 1866, to October 7th, 1867?

17. How much was to pay on taking up two notes, for $200 each, both due March 3d, 1868, and bearing interest at 6 %; one being dated Nov. 3d, 1867, and the other Feb. 3d, 1867?

SECTION 84.—1. What is the interest of $620, for 30 days, at 6 %?

MODEL. The interest for 60 days is $\frac{1}{100}$ of the principal, or $6.20; and for 30 days (which is $\frac{1}{2}$ of 60 days) it is $\frac{1}{2}$ of $6.20, or $3.10. *Ans.* $3.10.

2. Find the interest of $355, for 30 days, at 6 %.

3. Find the interest of $80, for 30 days, at 6 %.

4. Find the interest of $800, for 3 days, at 6 %.

5. What is the interest of $650, at 6 %, for 30 days? For 3 days? For 33 days?

6. What is the interest of $2040, at 6 %, for 90 days? For 3 days? For 93 days?

7. What is the amount of $18.60, at 6 %, for 60 days? For 3 days? For 63 days?

8. What is the interest of $1100, at 6 %, for 33 days? For 63 days? For 93 days?

9. What is the interest of $1200, for 1 year 10 months 18 days, at 6 %?

MODEL. The interest for 2 months is $\frac{1}{100}$ of the principal, or $12,—for 1 yr. 10 mo., or 22 months, it is 11 times $12, or $132,— and for 18 days (which is $\frac{18}{60}$ of 2 months) it is $\frac{18}{60}$, or $\frac{3}{10}$, of $12, which is $3.60. Hence for 1 yr. 10 mo. 18 da. it is $132 + $3.60, or $135.60. *Ans.* $135.60.

10. At 6 %, what is the interest of $336, for 2 years 12 days? For 1 yr. 7 mo. 10 da.?

11. At 6 %, what is the amount of $1500, for 10 months 15 days? For 1 yr. 11 mo. 5 da.?

12. At 6 %, what is the interest of $1800, for 4 months 5 days? For 1 yr. 3 mo. 25 da.?

13. At 6 %, what is the interest of $2000, for 8 months 21 days? For 3 yr. 4 mo. 8 da.?

14. At 6 %, what is the interest of $2400 for 6 days? At 6 %, the interest for 6 days is what part of the principal?

15. At 6 %, what is the interest of $8880 for 6 days? For 12 days? For 3 days? For 9 days?

16. Find the interest of $540, for 2 mo., at 7 %.

MODEL. At 6 % it would be $5.40; at 1 %, $\frac{1}{6}$ of $5.40, or 90c.; and at 7 %, it is 7 times 90c., or $6.30. *Ans.* $6.30.

Solve this Example according to the method shown in Section 80, and see whether the answers agree. Solve the following according to both methods.

17. Find the interest of $600, for 10 mo., at 7 %.
18. Find the interest of $9000, for 9 mo., at 5 %.
19. Find the amount of $420, for 18 mo., at 4 %.
20. Find the interest of $6500, for 14 mo., at 8 %.
21. Find the amount of $836, for 8 mo., at 7 %.

INTEREST. 135

SECTION 85.—1. What part of a year's interest is 1 month's interest?

2. Allowing 30 days to a month, what part of a year's interest is 1 day's interest?

3. How many days are there in a year? What part of 1 year's interest, then, would 1 day's interest exactly be?

4. By taking 1 day's interest as $\frac{1}{360}$ of 1 year's interest (in stead of $\frac{1}{365}$), do we take more or less than the exact amount? Is the difference great?

5. Why, do you suppose, are 30 days allowed to the month in computing interest?

6. Required the interest, by the shortest method,
    Of $70.50, at 6 %, for 60 days.
    Of $1040, at 6 %, for 90 days.
    Of $88.00, at 8 %, for 30 days.
    Of $3750, at 4 %, for 63 days.
    Of $225,   at 6 %, for 2 yr. 6 mo.
    Of $22.50, at 5 %, for 1 yr. 6 mo.

7. Required the amount, by the shortest method,
    Of $3.47,   at 10 %, for 10 years.
    Of $1400,   at 8 %, for 12 yr. 6 mo.
    Of $219.90, at 6 %, for 16 yr. 8 mo.
    Of $2000,   at 6 %, for 1 yr. 1 mo. 1 da.
    Of $860,   at 7 %, for 1 yr. 8 mo. 9 da.
    Of $900,   at 7 %, for 7 mo. 7 da.
    Of $400,   at 7 %, for 3 mo. 10 da.

8. Jan. 1st, A sells for B 6 bales of cotton, averaging 350 lb. each, at 25c. a pound. Jan. 21st, he remits to B $125. How much is due to B, Feb. 21st, interest being allowed at 6 % from the day of sale?

INTEREST.

SECTION 86.—1. At what rate will $200 yield $9 interest in 1 year?

MODEL. At 1%, $200 will yield $2 interest in 1 year; if it yields $9, the rate must be as many times 1% as $2 is contained times in $9, or 4½. *Ans.* 4½%.

2. At what rate must $650 be invested, to yield $39 interest annually? To yield $45.50?

3. At what rate of interest will $3000 produce $210 a year? At what rate will it produce $120?

4. If $2500 produces $200 interest in 12 months, what is the rate?

5. A person lends $1000, and at the end of a year receives for principal and interest $1065. What is the interest? What rate of interest does he receive?

6. If a person receives $481.50 for a loan of $450 for 1 year, what rate of interest does he get?

7. At what rate must £2000 be invested, to amount to £2060 in 1 year?

8. At what rate will $9000 yield $675 interest in 1 year 6 months?

MODEL. 1 yr. 6 mo. = 1½ or 3/2 yr. If the interest for 3/2 years is $675, for ½ year it is ⅓ of $675, or $225; and for 1 yr. it is twice $225, or $450. At 1%, $9000 will yield $90 in 1 year; if it yields $450, the rate must be as many times 1% as $90 is contained times in $450, or 5. *Ans.* 5%.

9. At what % will $800 produce $60 in 1 yr. 3 mo.?

10. At what % will $900 give $126 interest in 2 yr.?

11. At what rate will $240 produce $22.40 interest in 1 year 4 months?

12. At what rate will $606 produce $60.60 interest in 60 days?

INTEREST. 137

13. At what rate must $1000 be invested, to yield $50 interest in 10 mo.? To yield $200 in 4 yr.?

14. At what rate is $1100 invested, if it yields $38.50 interest semi-annually?

15. At what rate must $2500 be invested, to amount to $2850 in 2 years?

16. If the amount of $6700, for 1 yr. 8 mo., is $7370, what is the rate?

17. At what rate must $490 be put at interest, to amount to $627.20 in 3 yr. 6 mo.?

18. At what rate will the interest on any sum equal the principal in 10 years?

MODEL. The interest will equal the principal in 1 year at 100%, and in 10 years at $\frac{1}{10}$ of 100%, or 10%. *Ans.* 10%.

19. At what rate will $100 produce $100 interest in 4 years? In 6 years? In 20 years?

20. At what rate will the interest on a certain sum equal the principal in 12 years? In 5 years?

21. At what rate will any principal double itself in 2 years? In 8 years? In 15 years?

SECTION 87.—1. How long will it take $200, at 5%, to produce $22 interest?

MODEL. In 1 year, at 5%, $200 will produce $10, and to produce $22 will take it as many years as $10 is contained times in $22, or 2$\frac{1}{5}$. 2$\frac{1}{5}$ years = 2 yr. 2 mo. 12 da. *Ans.* 2 yr. 2 mo. 12 da.

2. How long will it take $650, at 6%, to yield $78 interest? How long, to yield $6.50?

3. How long will it take $100, at 7%, to yield $17.50 interest? How long, to yield $3.50?

4. How long will it take $3000, at 8 %, to produce $400 interest? To produce $260?

5. How long will it take $200, at 1 % a month, to produce $1? To produce $5?

6. At ½ % a month, in what time will $500 produce $15?

7. How long must $1600 be at interest, at 7 %, to yield $448 interest? To amount to $2048?

8. In what time will $1250, at 4 %, amount to $1312.50? To $1500?

9. At 7 %, in what time will the interest on any sum equal the principal?

MODEL. The interest, being 7 % of the principal in 1 year, will be 100 % of it, or equal to it, in as many years as 7 is contained times in 100, or 14²⁄₇. *Ans.* 14²⁄₇ years.

10. At 8 %, how long will it take $50 to yield $50 interest? How long, at 4 %? How long, at 5 %?

11. In what time will any principal double itself at 6 %? At 10 %? At 1 % a month?

12. In what time will any principal gain 50 % of itself at 7 %? At 5 %? At 4½ %?

13. At 6 %, how long will it take any principal to gain ½ of itself? To gain ¼ of itself?

SECTION 88.—1. What sum put at interest for 2 yr. 6 mo., at 6 %, will produce $225?

MODEL. The interest on $1, for 1 year, at 6 %, is 6c.; and for 2 yr. and 6 mo., 2½ times 6c., or 15c. The required principal is therefore as many times $1 as 15c. (which is ¹⁵⁄₁₀₀, or ³⁄₂₀, of $1) is contained times in $225, or 1500 times. *Ans.* $1500.

INTEREST.

2. What principal, in 9 months, at 8 %, will yield $72 interest?

3. What sum must be invested, at 7 %, to yield $115.50 annually?

4. A lady wishes to provide a semi-annual income of $200 for her son; how much must she invest in his name, at 5 %?

5. What principal invested at 6 % will yield $8.25 interest, in 1 year 10 months?

6. What sum put at interest at $6\frac{1}{2}$ % will produce $52 in 4 years?

7. What principal, at 1 % a month, will in 3 months 15 days yield $17.50 interest?

8. What principal, at $\frac{1}{2}$ % a month, will in 4 months amount to $408?

NOTE. At $\frac{1}{2}$ % a month, in 4 months $\frac{2}{100}$ or $\frac{1}{50}$ of the principal equals the interest. The principal being $\frac{50}{50}$ of itself, the amount must be $\frac{50}{50} + \frac{1}{50}$, or $\frac{51}{50}$, of the principal. The question therefore becomes, $408 is $\frac{51}{50}$ of how much?

9. What principal, at 7 %, will amount to $96.80 in 3 years?

10. A person, having borrowed some money for 1 yr. 3 mo., repays it with interest at 8 %, the amount being $352. What was the sum borrowed?

11. What principal will double itself in 10 years, at 10 %?

12. A person, having inherited 25 % of his father's property, put it out at interest at 6 %. With the amount, at the end of 2 years, he was able to buy $\frac{1}{2}$ of a mill valued at $4480. How much property was left by his father?

CHAPTER THIRTEENTH.

DISCOUNT.

SECTION 89.—1. What is **Discount?** *Ans.* An allowance made for the payment of money before it is due.

2. At $2\frac{1}{2}$ %, what will be the discount for cash on a bill of $800?

3. A person buys a bill of goods amounting to $1500, and can have 4 months' credit or a discount of 5 % for cash.* If he chooses the latter, what will be the discount, and how much cash will pay his bill?

4. A merchant buys $2400 worth of goods, 3 % being allowed for cash. How much cash will pay his bill, and what will the discount amount to?

5. Bought 50 pieces of cassimere, averaging 36 yards each, at $2 a yard, 5 % off for cash. For how much must a check be drawn, to pay the bill?

6. A person, having bought a bill of hardware, obtained a deduction of $22.50 by paying cash at a discount of $2\frac{1}{2}$ %; what was the amount of his bill?

7. If by paying cash a merchant gets a discount of $90 on a bill of $3000, what is the rate of discount?

8. A publisher sells a ten-dollar annual for 30 % less than its retail price, and then makes a discount of 5 % for cash. What must the purchaser sell it for, in order to make 20 %?

---

* This does not mean at the rate of 5 % a year, but 5 % on the face of the bill, without reference to time.

TRUE DISCOUNT.

SECTION 90.—1. How does True Discount differ from the discount treated of in the last Section? *Ans.* In computing True Discount, time is taken into account.

2. What is the **Present Worth** of a sum due at a future time without interest? *Ans.* Such a sum as put at interest for the given time will amount to the debt.

3. What is the **True Discount?** *Ans.* The difference between the Present Worth and the debt.

4. If I owe $53, due in 12 months without interest, what sum now paid would discharge the debt, money being worth 6 %, and what is the true discount?

MODEL. In 12 months, at 6 %, $1 would amount to $1.06,—that is, to $\frac{106}{100}$, or $\frac{53}{50}$, of itself. $53 is therefore $\frac{53}{50}$ of the present worth; $\frac{1}{50}$ is $\frac{1}{53}$ of $53, or $1; and $\frac{50}{50}$, or the present worth, is 50 times $1, or $50. The true discount is $53—$50, or $3. *Ans.* Present worth, $50; true discount, $3.

5. What is the present worth of $228, due 2 years hence, without interest, when money brings 7 %? What is the true discount?

6. At 5 %, what is the present worth of $90, due in 4 years? What is the true discount?

7. At 6 %, what is the present worth of $303, due in 60 days? What is the true discount?

8. At 8 %, what is the present worth of $1120, due 5 years hence?

9. A merchant, having bought 40 clocks, at $25.50 apiece, on 4 months' credit, afterwards gets a discount of 3 % for cash. By how much does this discount exceed the true discount, at 6 %?

BANK DISCOUNT.

SECTION 91.—1. What is **Bank Discount?** *Ans.* An allowance made to a bank for cashing a note before it is due.

2. How is Bank Discount computed? *Ans.* At a certain % on the face of the note, which the bank retains, paying over the balance to the owner. The balance thus paid over is called the **Proceeds.**

3. A bank discounts a note for $200, which will mature in 3 mo., at 7 %; what is the discount, and what are the proceeds?

MODEL. At 7 %, the bank discount on $200, for 1 year, would be $14, and for 3 months it is ¼ of $14, or $3.50. The proceeds are $200—$3.50, or $196.50. *Ans.* Discount, $3.50; proceeds, $196.50.

4. At 6 %, what is the bank discount on a note for $1000, to mature in 63 days? On a note for $1200, to mature in 33 days?

5. Required the proceeds of a note for $700, maturing Sept. 7th, and discounted the 7th of the previous June, at 6 %.

6. A person invested the proceeds of a note for $1200, discounted for 4 months, at 6 %, in flour at $8 a barrel. How many barrels did he buy?

7. What is the bank discount on a note for $900, to run 30 days, at 7 %?

8. The bank discount, for 2 months, on a $300 note, was $3.50; what was the rate?

9. The proceeds of a $600 note, discounted for 90 days, were $591; what was the rate?

10. The proceeds of a note for $1000, discounted at 6 %, were sufficient to pay for 98 acres of land, at $10 an acre. How long had the note to run?

CHAPTER FOURTEENTH.

STOCKS.—U. S. SECURITIES.

SECTION 92.—1. What is a **Bond**? *Ans.* A Bond is a written instrument by which one party binds himself to pay another a certain sum.

2. What is meant by **Stocks?** *Ans. Stocks* is a general term applied to Government or State bonds, and the capital of incorporated companies.

3. How is stock divided? *Ans.* Into shares, generally of $100 each.

4. When is a stock **at par?** *Ans.* When it sells for its nominal value. If it sells for more than its nominal value, it is **above par**, or **at a premium**; and if less, **below par**, or **at a discount**.

5. What are 100 shares of Erie Railroad stock worth, at $69\frac{1}{2}$?

MODEL. At $69\frac{1}{2}$, one share ($100) is worth $69.50; and 100 shares are worth 100 times $69.50, or $6950. *Ans.* $6950.

NOTE. Take $100 for a share, unless it is otherwise specified.

6. What are 100 shares of Chicago and Northwestern worth, at 67?

7. What is the value of 50 shares of N. Y. Central, at 125 (25 % above par)?

8. How much are 100 shares of Pacific Mail worth, at 8 % below par?

9. If N. J. Central is selling at a premium of $16\frac{1}{4}$ %, how much are 80 shares worth?

10. If 100 shares of Milwaukee and St. Paul are worth $6500, how much below par is the stock?

SECTION 93.—1. What is a **Stock-broker?** *Ans.* One who buys and sells stock for others, at a charge usually of $\frac{1}{4}$ of 1 % *on the par value of the stock bought or sold.*

2. What is the brokerage on 100 shares of Hudson River R. R. stock, bought at 137?

MODEL. $\frac{1}{4}$ % on 1 share of $100 is $\frac{1}{4}$ of a dollar, and on 100 shares it is 100 times $\frac{1}{4}$, or $25. *Ans.* $25.

NOTE. Reckon brokerage at $\frac{1}{4}$ %, unless it is otherwise specified.

3. What is the brokerage on 300 shares of Cleveland and Toledo, sold at 106?

4. What is the brokerage for buying and selling 150 shares of Great Western?

5. For what % advance must a person sell stock, to cover the brokerage for buying and selling, and make 1 %?

6. What is the cost, including brokerage, of 100 shares of Western Union Telegraph stock, at 40?

7. How much cash should be given with 100 fifty-dollar shares of a mining company, selling at 25, for 50 shares of Ohio and Mississippi, worth 30?

8. What is the profit, over and above brokerage, on 100 shares of stock bought at 94, and sold at 97?

9. What is the loss, including brokerage, on 50 shares of stock bought for 67, and sold for 60?

10. A person has bought a hundred shares of stock at 99; to make $500 on his purchase, what must he sell it for?

11. A large dealer in stocks arranges with a broker to buy and sell for him at $\frac{1}{8}$ %. What will the brokerage be on 300 fifty-dollar shares?

12. Fifty shares of bank stock were bought at par, and sold at 103. What was the profit, brokerage being paid on each transaction?

13. A person realized $200 (leaving brokerage out of account) by selling 100 shares of stock at 75. At what rate did he buy them?

14. By selling 200 shares of stock at $81\tfrac{1}{2}$, a person lost $450. What did they cost him?

15. How many shares 20 % below par can be bought for $6000, leaving brokerage out of account?

SECTION 94.—1. What is a **Dividend?** *Ans.* A sum paid from the earnings of a company to those who hold its stock.

2. How is a dividend reckoned? *Ans.* At a certain % *on the par value* of the stock.

3. A bank declares a dividend of 7 %. What will a person receive who holds stock worth at par $8000?

4. A railroad company having declared a dividend of 3 %, how much will a party who owns 50 shares receive?

5. A ferry company pays a dividend of 1 % a month. How much will a person who owns 300 twenty-five-dollar shares receive in a year?

6. A person who bought some stock at 50, receives from it a yearly dividend of 6 %. What per cent. does he get on his investment?

MODEL. Each $100 share draws a dividend of $6, and, being bought at 50, cost $50. He therefore makes $\tfrac{6}{50}$, or $\tfrac{12}{100}$. *Ans.* 12 %.

7. If a person buys some stock at 96, and receives from it yearly dividends of 6 %, what per cent. does he get on the investment?

8. What per cent. on the investment will be realized from stock bought at 60, and paying a semi-annual dividend of $2\frac{1}{2}$ %?

9. If 50 shares of stock draw a dividend of $400, what is the rate of dividend? If said stock was bought at 160, what % does it pay on the investment?

10. At what rate must bonds that pay 6 % annually be bought, to yield 8 % on the investment?

MODEL. Each $100 of bonds pays $6, which must be 8 % of the required rate. If $6 is $\frac{8}{100}$, or $\frac{2}{25}$, of the required rate, $\frac{1}{25}$ is $\frac{1}{2}$ of $6, or $3; and $\frac{25}{25}$ is 25 times $3, or $75. *Ans.* 75.

11. At what rate must 5 % bonds be purchased, to pay 8 % on the investment?

12. A person receiving $300 dividend on 100 shares of stock, gets 6 % on his investment. How much below par did he buy the stock?

SECTION 95.—1. When gold is at 140, how much in greenbacks will it take to buy $250 in gold?

MODEL. At 140, $100 in gold will cost $140 in greenbacks; and $250 in gold will cost $2\frac{1}{2}$ times $140, or $350. *Ans.* $350.

NOTE. Since 1861, gold and silver have commanded a premium; that is, $1 in gold or silver has been worth more than $1 in currency.

2. A merchant, having to pay his duties in gold, needs for this purpose $300. What will it cost in currency, when gold is at 114?

3. Having $450 in gold, a person sold it at 136½. How much more would he have got for it, had he waited till the next week, when gold rose to 141¼?

4. What must be paid for 10 gold eagles, when gold is at a premium of 12¾?

5. Bought 10 Swiss watches, at $100 each in gold. What sum in currency will pay the bill, gold being 138?

6. A person, having bought some jewelry, has the choice of paying $6000 in gold, or $8420 in current funds. Gold being 140, which had he better do?

7. When gold is 150, how much gold can be bought for $600 in currency?

8. Having $4200 in currency, a person invests it in gold at 110, which he sells the next day at 111; how much does he make?

9. A merchant changes $650 currency into gold, to pay the duty on some drygoods. If gold is worth 130, how much does he buy?

10. A person who had received a 3% dividend on 90 shares of railroad stock, bought half eagles with it. How many did he buy, if gold was 135?

11. How much more gold can be bought for $6000, when gold is 120, than when it is 150?

12. When gold stood at 180, how much currency was $900 in gold worth? How much gold was $900 in currency worth?

13. With gold at 160, how much more, in gold, is a person worth who has 30 double-eagles, than one who has $800 in greenbacks?

14. Twenty eagles were bought for $280 in currency; how did gold stand?

U. S. SECURITIES.

SECTION 96.—1. What is meant by **Government Bonds** or **U. S. Securities?** *Ans.* Bonds issued by the United States Government.

2. What are Five-twenties? *Ans.* U. S. Bonds, bearing interest at 6 % in gold, payable in not less than five or more than twenty years from their date.

3. What are Ten-forties? *Ans.* U. S. Bonds, bearing interest at 5 % in gold, payable in not less than ten or more than forty years from their date.

4. What are U. S. Currency 6's? *Ans.* U. S. Bonds payable in 1895 and thereafter, bearing interest at 6% in currency.

5. What cost $1000 of five-twenties, at 106 ?

6. At 103½, what are $5000 of 10-40's worth ?

7. A person who had 100 shares of Erie Railroad stock, sold them at 70, paying brokerage, and bought 10-40's to the amount of $6000, at 103 ; how much of the proceeds of his stock had he left ?

8. A person who had $2200 in currency, bought two 5-20 bonds of $1000 each, at 108. How much currency had he left ?

9. If I get a note for $2500 discounted at a bank, for 60 days, at 6 %, how much more than the proceeds of this note will I need, to buy $2500 in U. S. Currency 6's, at 115 ?

10. What amount of 5-20's, at 110, can be bought for $3300 ?

11. What amount of 10-40's, at 104, will $5200 buy ?

12. A person sold $5000 in gold, at 115, and with the proceeds bought six 5-20 bonds of $1000 each, at 110. How much had he left ?

U. S. SECURITIES.

SECTION 97.—1. What income in gold will a person annually receive from $2000 of five-twenties? From $8000 of ten-forties?

2. What income in gold will a person receive half-yearly from $3500 of ten-forties? From $7500 of five-twenties?

3. When gold is at 138, what is the semi-annual income in currency from $5000 of five-twenties?

MODEL. The income for 1 year is $\frac{6}{100}$, and for half a year $\frac{3}{100}$, of $5000,—which is $150, in gold. If gold is at 138, $100 in gold is worth $138 in currency; and $150 in gold is worth $1\frac{1}{2}$ times $138 in currency, or $207. *Ans.* $207.

4. When gold is 115, what is the value in currency of one year's interest on two five-twenty bonds of $1000 each?

5. When gold is 140, what is the value in currency of six months' interest on six ten-forty bonds of $500 each?

6. What is the yearly income from $1000 of 6's?

7. What amount of five-twenties will yield a yearly income of $540 in gold?

8. How much must be invested in ten-forties, to yield $100 in gold every six months?

9. A person had equal amounts invested in five-twenties and ten-forties. If his income from these bonds was $330 a year in gold, what amount of each kind had he?

10. B has equal amounts invested in 5-20's, 10-40's, and U. S. Currency 6's. If his 5-20's yield $60 a year in gold, what will be his whole income from these bonds in currency, when gold is 112?

CHAPTER FIFTEENTH.

MISCELLANEOUS EXAMPLES.

SECTION 98.—1. What is the cost of a draft on Charleston for $2250, at 1% premium?

2. What is the cost of a draft on Cincinnati for $1000, at ½% discount?

3. If a franc is worth $19\frac{3}{10}$c., what is the value of 1000 francs in U. S. currency?

4. The pound sterling of Great Britain is worth 4.86\frac{65}{100}$, and the Canada dollar is worth $1, in U. S. gold. How many Canada dollars are equivalent to 500 pounds sterling?

5. How many Canada dollars are 100 eagles worth?

6. A grocer mixes 100 lb. of coffee, at 24c., with 20 lb., at 15c.; what is the mixture worth per lb.?

MODEL. 100 lb., at 24c. a pound, are worth $24. 20 lb., at 15c., are worth $3. The whole, 120 lb., is therefore worth $24+$3, or $27; and 1 lb. is worth $\frac{1}{120}$ of $27, or 22½c. *Ans.* 22½c.

7. If 10 gall. of brandy, at $4 a gallon, were mixed with 5 gall., at $10 a gallon, how much a gallon was the mixture worth?

8. If a train of cars goes 20 miles an hour for 2 hours, and then 30 miles an hour for 3 hours, what is its average rate for the whole time?

9. If a boat runs 30 miles in 2 h., and then 15 miles in 1½ h., what is its average speed?

10. A grocer mixes 20 lb. of tea that cost 70c. a lb. with 20 lb. that cost $1.10 a lb. At what price per lb. must he sell the mixture, to make 20%?

11. Two thousand bushels of wheat, worth $2.50 a bushel, were mixed with 1000 bu., worth $1.90. How much a bushel was the mixture worth?

12. An equal number of geese, chickens, and turkeys, were sold for $15. Each goose brought 90c., each chicken 50c., and each turkey $\frac{11}{14}$ of the cost of a goose and chicken; how many of each were sold?

13. A person bought some property for $2000 on the 10th of Jan., and on the 10th of May following sold it for $2200. What % did he make, and how much better off was he than if he had loaned his money at 7 % for the same time?

14. 4 times $\frac{5}{6}$ of 36 is $\frac{1}{3}$ of what number?

SECTION 99.—1. How much will a pile of wood, 16 feet long, 4 feet wide, and 8 feet high, cost, at $6 a cord?

2. If A can do a certain piece of work in $\frac{3}{4}$ of an hour, and B can do it in $\frac{5}{6}$ of an hour, how long will it take both to complete the job after B has been working 20 minutes?

3. The sum of two numbers which are to each other as 5 to 4, is 36; what are the numbers?

4. The difference between two numbers which are to each other as 5 to 9, is 24; what are the numbers?

5. A rockaway was sold for $220, at a profit of 10 %; at what price would it have brought a profit of 25 %?

6. What principal, at 4 %, for 2 years, will yield as much interest as $200, at 6 %, for 3 years?

7. A commission-merchant sold a consignment of goods for $6000. If he paid $450 expenses, and his commission was 2½ % on the sales, how much should he remit to the consignor?

8. A person sold two houses for $3960 each, making 10 % on one, and losing 10 % on the other. Taking both sales into account, what was his gain or loss?

9. What principal will, in 2 years, at 7 %, amount to a sum sufficient to buy 76 acres of land, at $30 an acre?

10. The owner of a farm let it out to a party to work on shares, allowing him 40 % of all he raised. How many bushels of potatoes were raised, if the owner's share was 480 bushels?

11. A person who received 75 % of the rent of a hotel, with 50 % of his income for 1 year from this source bought ⅓ of a mill. If the hotel rented for $4000 a year, how much was the mill worth?

12. Four-fifths of a vessel was sold for $5760, at a loss of 10 %. How much would the whole vessel have had to sell for, to bring a profit of 10 %?

13. A and B had different amounts of five-twenties, A's being to B's as 3 to 5. When gold was at 150, the two realized $720 yearly interest in currency from these bonds; what amount had each?

14. A man whose money was invested at 6 %, succeeded in changing the investment so as to get 7 %, and found that it made a difference of $37.50 in his semi-annual income. How much money had he?

15. At how much a pound must guano be sold, to make 20 %, if it cost $85 a ton?

SECTION 100.—1. A's money was invested at 5 %, B's at 6 %, and C's at 7 %. The sums invested were to each other as 1, 2, and 3. If their yearly incomes from these investments were together $380, how much had each invested?

2. A certain principal, at 5 %, amounted in a year to $8.75 less than it would have amounted to in a year and a half. What was the principal?

3. Three-fourths of John's age is 1 year less than Henry's, and 1 year more than Daniel's. The sum of their ages being 50 years, how old is each?

4. A person left some money to be divided between three of his relatives in the proportion of $\frac{1}{2}$, $\frac{1}{3}$, and $\frac{1}{4}$. The money was invested at 6 %, and brought $234 yearly interest. How much of the principal should each receive?

5. D sold a horse to E at a profit of 50 %; E sold him to F at a loss of 50 %. If F gave $150 for the horse, what did D give for him?

6. At $7 a cord, what will a pile of wood, 32 ft. long, 4 ft. wide, and 6 ft. high, cost?

7. How much water must be added to 21 gal. of alcohol, worth $4 a gallon, to make it worth but $3 a gallon?

8. Four parties, having speculated with $6000, realized a profit of 20 %. A contributed $\frac{1}{6}$ of the capital, B $\frac{1}{3}$, C $\frac{5}{12}$, and D $\frac{1}{12}$. The transaction being finished, how much should each get for his share of the profit and capital?

9. At $3.50 a rod, what will be the expense of fencing a field 3 rods square?

10. On what day did a note for $800 mature, which was discounted August 10th, at 6%, for $2?

11. What part of a plot 1 rod square is a bed ½ of a rod square?

12. How many pounds of coffee worth 12c. a lb., must be mixed with 9 lb. worth 20c. a lb., to make the mixture worth 15c. a lb.?

MODEL. On each pound put in at 20c., to be sold at 15c., there is a loss of 5c.; and on 9 lb. there will be a loss of 9 times 5c., or 45c. On each pound put in at 12c., to be sold at 15c., there will be a gain of 3c.; and, to balance the loss of 45c., there must be as many pounds at 12c. put in as 3c. is contained times in 45c., or 15. *Ans.* 15 lb.

13. How many pounds of tea at $1.25 a pound, must be mixed with 10 lb. at 80c. a pound, to make the mixture worth $1 a pound?

SECTION 101.—1. When gold is quoted at 150, what is the value in gold of $1 in currency?

2. How many hours will a person save in the three summer months by sleeping but 7 hours daily in stead of 8?

3. How many gallons of whiskey, at $2 a gallon, must be mixed with 12 gal. worth $3.50 a gallon, to make the mixture worth $3 a gallon?

4. Two clocks were sold for $60 apiece, at a loss of 20% on one, and a profit of 20% on the other. Taking both sales into account, was there a gain or loss, and if either how much?

5. ⅝ of 1¾ is ⅜ of how many times 5% of 400?

MISCELLANEOUS EXAMPLES.

6. What o'clock is it, if the time past 12 is $\frac{1}{2}$ of the time from now till 1?

7. How many hours will you lose in a leap-year, if you are idle 5 minutes every day?

8. In going $\frac{1}{4}$ of a mile, how many more times will a wheel turn that is 12 feet in circumference, than one that is 15 feet?

9. If a certain principal, at 6%, produces $135 in 2 yr. 3 mo., how much interest will the same principal produce in 1 yr. 6 mo., at 7%?

10. The pipe A can fill a cistern in $\frac{2}{3}$ of the time that the pipe B can fill it. If both can fill it in 20 minutes, how many times can the pipe A alone fill it in 1 hour?

11. How many bushels of wheat, at $2.50 a bushel, can be bought with the proceeds of a note for $1500, discounted at a bank, for 2 mo., at 6%?

12. Two oblong fields contain the same number of square rods. The first is 50 rods long and 6 rods wide. The second is 20 rods long; how wide is it?

13. Ida is now half as old as Jane, but in 6 years she will be $\frac{3}{4}$ as old as Jane will then be; what is the age of each?

MODEL. Ida is now half as old as Jane. To maintain the same ratio, she would have to live only 3 years to Jane's 6; and therefore at the end of 6 years she will be half as old as Jane will then be, and 6—3, or 3, years more. But by the conditions she will then be $\frac{3}{4}$ as old as Jane will be; hence 3 years must be the difference between $\frac{1}{2}$ and $\frac{3}{4}$ of Jane's age six years hence. Jane will therefore be 12, and Ida $\frac{3}{4}$ of 12, or 9; and the present age of each will be 6 years less. *Ans.* Jane, 6; Ida, 3.

SECTION 102.—1. Susan is now 3 times as old as Sarah; four years hence, her age will be twice Sarah's. How old is each?

2. Anna is 8 years old. Ella's age equals Anna's increased by $\frac{1}{3}$ of Jacob's; and Jacob's equals Ella's increased by $1\frac{1}{2}$ times Anna's. How old are Ella and Jacob?

MODEL. Anna being 8 years old, Ella's age equals $\frac{1}{3}$ of Jacob's + 8 years, and Jacob's equals Ella's + 12 years. Hence $\frac{1}{3}$ of Jacob's age is $\frac{1}{3}$ of Ella's + 4 years; and Ella's age must equal $\frac{1}{3}$ of her age + 4 years + 8 years. 12 years, therefore, equals the difference between Ella's age and $\frac{1}{3}$ of her age, or is $\frac{2}{3}$ of Ella's age. Hence Ella is 18; and Jacob's age, being equal to Ella's increased by $1\frac{1}{2}$ times Anna's age, or 12 years, is 30. *Ans.* Ella, 18; Jacob, 30.

3. A person, buying a horse, wagon, and harness, paid for the harness $40; for the wagon as much as for the harness and $\frac{1}{3}$ of the cost of the horse; and for the horse as much as for the wagon and twice the cost of the harness. What was the cost of the whole?

4. There are three poles, the first of which is as long as the other two; the second is 6 feet; and the third is as long as $\frac{1}{2}$ the first increased by $\frac{2}{3}$ of the second. What is the length of the first and third?

5. How many acres in an oblong garden, 20 rods long by 10 rods wide? How many rods of fence will be needed to enclose it?

6. If $\frac{1}{8}$ of a piece of work is done by 2 men in 6 days, how many men will it take to do what remains in 4 days?

7. What is the amount of $600, for 2 years 3 months 15 days, at 6 %?

8. From what number must $\frac{2}{5}$ of 40 be taken 3 times, to leave 3?

9. The product of two numbers is 36; if one of the factors is $\frac{1}{6}$ of 54, the other is $\frac{4}{5}$ of what number?

10. Some apples were bought at the rate of 3 for 2c., and sold at the rate of 2 for 3c. If the profit was 50c., how many apples were there? What was the per cent. of profit?

11. A wall containing 180 square feet is 18 feet long; another wall, equally high, is 12 feet long. What is the area of the second wall?

12. If 4 horses can remove $\frac{1}{5}$ of a heap of stone in $\frac{5}{8}$ of a day, how many horses will be needed to remove the whole heap in half a day?

SECTION 103.—1. Divide $1 between three persons, so that the first may have $$\frac{1}{10}$ more than the third, and $$\frac{7}{20}$ less than the second.

2. A publisher takes off $33\frac{1}{3}$ % from his retail prices for a wholesale customer; what % will the purchaser make, if he sells at the publisher's retail prices?

3. A and B are to dig a cellar for $28. When $\frac{1}{5}$ of the work is done A is taken sick, and B has to finish it. Divide the $28 fairly between them.

4. C and D agreed to cut some wood for $60. When the work was partly done, D was taken sick and received only $10. What part of the job was finished when D quit work?

5. $\frac{2}{7}$ of $1\frac{4}{15}$ is 8 times $\frac{1}{4}$ of how many times $\frac{1}{30}$?

6. What % on the investment is a 2 % semi-annual dividend on stock bought at 60 % below par?

7. How many boxes, 1 ft. long, 1 ft. wide, and 6 in. high, can be packed in a space 4 ft. each way?

8. A lady bought some eggs at the rate of 3 for 5 cents, and had 10c. left. Had she given 25c. a dozen, she would have needed 5c. more to pay for them. How many did she buy?

9. How many yards of silk $\frac{3}{4}$ of a yd. wide will line 4 satin damask curtains, each 12 feet long by 6 feet wide?

10. A person bought 9 tons of coal for $88; partly soft coal, for which he gave $12 a ton, and partly hard, which cost $7. How many tons of each did he buy?

11. Two blocks of stone contain the same number of cubic feet. They are both 6 ft. long; the first is 4 ft. wide and 3 ft. high; the second is $1\frac{1}{2}$ ft. wide,— how high is it?

12. One number is $1\frac{2}{3}$ times another; if the difference between them is 10, what are the numbers?

13. In what time will $48.77 amount to $97.54, at $6\frac{1}{2}$ %? At what per cent. will it amount to $97.54 in 12 yr. 6 mo.?

14. What o'clock is it, if the time past 12 is $\frac{1}{3}$ of the time past 11?

MODEL. The time past 11 is $\frac{3}{3}$ of itself; and, since the time past 12 is $\frac{1}{3}$ of the time past 11, the time between 11 and 12, or 60 min., must be the difference between $\frac{3}{3}$ and $\frac{1}{3}$, or $\frac{2}{3}$, of the time past 11. If 60 minutes are $\frac{2}{3}$ of the time past 11, $\frac{1}{3}$ must be $\frac{1}{2}$ of 60 min., or 30 min.; and $\frac{3}{3}$ is 3 times 30 minutes, or 90 minutes. *Ans.* 90 minutes past 11,—that is, half past twelve.

SECTION 104.

1. A stage is going at the rate of 10 miles in 1⅓ h.; a train of cars is coming up behind at the rate of 10 miles in 22 min. If the train overtakes the stage in 88 minutes, how far was it behind the stage at first?

2. What o'clock is it, if the time past 3 is ¼ of the time past 1?

3. What o'clock is it, if the time past 4 is ⅓ of the time from now to 5?

4. A crown is worth 5s. How many more books, at half a crown each, can be bought for 10 guineas than for £10?

5. A person bought a section, or square mile of land, for $800. He divided it into forty-acre farms, which he sold for $60 each. Did he gain or lose, and what %?

6. After running at the rate of 20 miles an hour for 20 minutes, a train increased its speed 50%; how far did it run in all during the first hour?

7. Three brothers have money at interest, at 6%, from which they realize $900 yearly. A has ⅔ of the principal, and of the rest B has 4 times as much as C. How much has each?

8. An importer has to pay a duty of 60% in gold on silk invoiced at $7000; when gold is at a premium of 40%, what will the duty cost in currency?

9. A garrison of 100 men have food enough for 60 days, allowing each man 3 lb. a day. After 30 days, 50 more men join them, and their daily allowance is diminished half a pound. How long will their supplies then last?

10. A owes B $400, payable in 2 months, and $200, payable in 5 months. At what time should the whole be paid, so that neither party may gain or lose?

Model. The use of $400 for 2 mo. is equivalent to the use of $1 for 400 times 2 mo., or 800 mo. The use of $200 for 5 mo. is equivalent to the use of $1 for 200 times 5 mo., or 1000 mo. Hence A is entitled to the use of $1 for 800 mo. + 1000 mo., or 1800 mo.—and to the use of the whole money owed ($400+$200, or $600) for $\frac{1}{600}$ of 1800 mo., or 3 mo. *Ans.* 3 months.

11. A merchant sells a customer $1000 worth of goods on 3 months' credit, and $500 worth on 6 months'. The purchaser wishes to make one payment of the whole; when should it be made?

12. A person buys $1800 worth of goods; $\frac{1}{2}$ of the bill, on 1 month's credit; $\frac{1}{4}$, on 2 months'; and the rest, on 4 months'. In what time may he equitably make one payment of the whole?

13. When should a party discharge a debt of $600 in one payment, if 75% of it is due in 10 days, and the rest in 30 days?

14. A trader bought a bill for $1000,—half for cash, and half on 3 months' credit. If he gives a note for the whole, in what time should it mature?

Model. Half the bill being for cash, the trader is entitled only to the use of $500 for 3 mo.; which is equivalent to the use of $1 for 1500 mo., or $1000 (the amount of the bill) for $\frac{1}{1000}$ of 1500 mo., or 1½ mo. *Ans.* 1½ mo.

15. When should a bill for $4000, half cash, and half due in 1 mo., be discharged in one payment?

16. What is the process treated of on this page called? *Ans.* **Equation of Payments.**

SECTION 105.—1. A lent B $500 for 6 months; how long should B lend A $1200, that neither may gain or lose interest?

2. Five times a number exceeds 3 times $16\frac{2}{3}\%$ of the same number by 108; what is the number?

3. A person, being asked the hour, said that the time that would elapse before midnight was $\frac{3}{4}$ of the time past noon; what o'clock was it?

4. The shell of a certain cocoa-nut weighs 5 oz.; the milk weighs $\frac{1}{2}$ as much as the shell added to $\frac{1}{4}$ the weight of the kernel; the kernel weighs as much as the shell and half the milk. What is the weight of the whole cocoa-nut?

5. If Isaac had as many more books as he now has, half as many more, and 40 volumes besides, he would have twice five score books; how many has he?

6. A boy having taken 44 steps, a man starts off after him, taking 4 steps to the boy's 3. If 3 of the man's steps equal 5 of the boy's, how many steps will the man take before he comes up to the boy?

7. Two workmen engage to do a job for $51. The first, being the better workman, is to have $9 as often as the second has $8. They finish the work in 8 days; what wages per day does each make?

8. If 2 bu. 2 pk. of seed is allowed to an acre, how much will the seed for 2 A. 2 R. cost, at $2.50 a bushel?

9. How much rum, at $2 a gallon, must be mixed with 3 gallons at $4, to make the whole worth $2.50 a gallon?

10. At 7%, what is the present worth of 80% of $8025, due 1 year hence without interest?

11. What is P's income tax, if he has an income of $2400 over and above what is exempt, the rate being 5 %, and has to pay on 35 oz. of silver at 5c. an oz.?

12. A person insured his life for $3000 at the rate of $3.10 on $100. After paying 4 premiums, he died; how much more did his family receive than was paid out for premiums?

13. A dishonest milkman adds a quart of water to every gallon of milk, and then sells the mixture at 10 % more a quart than he paid for the pure milk. What % profit on the whole does he make?

14. A can do a piece of work in 3 days, B in 4 days, C in 5 days. If they do the job together for $9.40, and each is paid according to the work he does, how much should each get?

SECTION 106.—1. When gold is 130, is it better to lend money at 7 % or to buy five-twenties at 106?

MODEL. $100 of five-twenties, at 106, will cost $106 in currency. The yearly interest on $100 of five-twenties is $6 in gold, or, when gold is 130, $7.80 in currency. The yearly interest on $106 (the cost of $100 of five-twenties), at 7 %, would be $7.42. *Ans.* Five-twenties would pay the better interest.

2. When gold is at 140, which is the better, an investment on bond and mortgage at 6 %, or in five-twenties at 110?

3. With gold at 150, is it better to lend money at 1½ % a month or to buy ten-forties at 105?

4. Is it better to buy State bonds paying 6 %, at par, or 5 % bonds at 90?

MISCELLANEOUS EXAMPLES.

5. A pole is fixed in the bottom of a river. Three feet are in the air; the part in the water is 3 times as long as that in the mud; and the part in the mud is $\frac{4}{15}$ of the rest of the pole. How long is the pole?

6. Divide 45 into four numbers, each of which shall be half of the next greater.

7. What fraction of a surface 1 ft. 8 in. square is 8 square inches?

8. Two boys buy 40 pears each, at 2c. apiece. One sells his at 30c. a dozen. The other sells $\frac{1}{4}$ of his at cost, and the rest at 3c. apiece. What % does each make?

9. How high must a wood-cutter make a pile of wood which is 4 ft. wide and 36 ft. long, to contain 9 cords?

10. Arthur gave half his marbles to Hugh, who gave $\frac{1}{4}$ of what he thus received to Alfred. Alfred, after winning 6 more, gave all he had to Arthur, who then found that he had $\frac{3}{4}$ of his original number. How many had Arthur at first?

11. Divide 160 into four numbers, each of which shall be 3 times as great as the next smaller.

12. An agent received $3150, from which he was to take his own commission (5 % on the money invested), and with the rest buy land, at $2 an acre. How many acres did he buy?

13. If 5 men can do as much work as 8 women, and 3 women can do as much as 5 boys, and 2 boys can do as much as 3 girls, how many girls will it take to do as much as 4 men?

14. 22 % of 150 is what % of the difference between $\frac{7}{10}$ of 200 and $\frac{4\,7}{1\,0}$ of 500?

MISCELLANEOUS EXAMPLES.

SECTION 107.—1. If I ask for a farm 20% more than it cost, but fall 10% on my asking price, what % do I make by the sale?

2. Bought 5 yd. of cloth, at $4.50 a yard; 2 pair of gloves, at $1.25 a pair; 3 dress patterns, at $5 each; and ½ dozen handkerchiefs at $13 a dozen. What did the bill amount to?

3. A tenant has the choice of paying $500 rent in advance, or $550 at the end of the year. If he can borrow the money to make the advance payment with, at 6%, which is it better for him to do?

4. Three numbers multiplied together give a product of 60. Two of the factors are to each other as $\frac{1}{3}$ to $\frac{5}{6}$, and the third is $1\frac{1}{2}$. What are the numbers?

5. At $5 a thousand, what is the cost of 24 packs of envelopes, containing 25 each?

6. If a sovereign (the coin that represents £1 sterling) is worth $4.86 in specie, what were 10 sovereigns worth in currency when gold was at 150?

7. From a pile of wood, 40 ft. long, 4 ft. wide, and 6 ft. high, was sold $27 worth, at $6 a cord. How much was what remained worth, at $5½ a cord?

8. For a cabin passage to Liverpool, one steamer charges $120 in gold, and another $150 in currency. What premium must gold bring, to make these prices equal?

9. The interest on $\frac{1}{8}$ of A's fortune and $\frac{3}{16}$ of B's fortune, for 1 year, at 7%, is $140. If A's fortune is $\frac{4}{5}$ of B's, what is the fortune of each?

10. Mary is now 8 yr. older than Ruth; two years ago, she was twice Ruth's age. How old is each?

11. Augustus earns $15 a week, and determines to save enough to present his mother $50 on Christmas. If it is ten weeks to Christmas, what per cent. of his salary must he save?

12. A dog overtook a fox after running half a mile. Four fifths of the distance the fox ran after the dog started, was 8 rods less than 6 times the start he had? How many rods' start had the fox?

13. Fifty men have provisions to last them 60 days, at a certain rate of supply. Ten more men coming, and the daily supply being made $\frac{1}{4}$ less than it was before, how long will the provisions last?

14. $\frac{3}{4}$ of 8 times $\frac{2}{7}$ of 21 is how many times $\frac{2}{3}$ of $13\frac{1}{2}$?

SECTION 108.—1. Twelve bottles, holding 1 pt. 3 gi. each, being filled from a cask of wine containing 30 gallons, how much is the rest of the wine worth, at $1.50 a quart?

2. Two numbers are to each other as 5 to 6, and $\frac{1}{8}$ of their difference is 1. What are the numbers?

3. If a certain number increased by 2 is multiplied by 5, and the product divided by 3, we shall have 15. What is the number?

NOTE. What number divided by 3 equals 15?
What number multiplied by 5 equals 45?
What number increased by 2 equals 9?

4. If a certain number diminished by 4 is divided by $\frac{1}{5}$ of 20, and the result multiplied by 5, we shall have $2\frac{1}{2}$. What is the number?

5. A farmer sold 25 % of his potatoes, and then 25 % of what remained. 25 % of what were then left rotted, and he had 135 bushels of good potatoes on hand. How many had he at first?

6. If you walk at the rate of a mile in 20 minutes, how many miles can you walk in the time that you would save by sleeping half an hour less every day during the month of July?

7. What will it cost to insure a house worth $6000, and furniture in it worth $3000, for two thirds of their value, at $\frac{3}{10}$ of 1 %, 5 % being deducted from the premium for cash?

8. A bankrupt failed, having $7500 liabilities, and $1500 assets. How much could he pay on the dollar? What was B's loss, whom he owed for 12 pieces of muslin, containing 40 yd. each, at 25c. a yd.?

9. What cost 2 bu. 3 pk. 6 qt. of oats, at 75c. a bu.?

10. If a grocer in 10 days sells 12 cwt. 20 lb. of sugar, what do his daily sales of sugar bring in, on an average, at 12c. a pound?

11. If 2 hands are $\frac{2}{3}$ of a span, if a span is $\frac{1}{2}$ of a cubit, and a cubit $\frac{1}{4}$ of a fathom, how many feet are 10 fathoms equal to, the hand being 4 inches?

12. How many cubes $\frac{1}{4}$ ft. long, wide, and high, can be packed in a space 1 ft. long, wide, and high?

13. If there are 3 miles in a league, and a boat has to go 10 leagues, what part of the distance has she made when she has gone 160 rods?

14. From 2 qt. of seed was raised 1$\frac{1}{4}$ bu. of grain; what per cent. of the seed was the crop?

15. $\frac{5}{6}$ of 18 is how many tenths of $\frac{1}{2}$ of 60?

MISCELLANEOUS EXAMPLES.

SECTION 109.—1. At what time between 12 and 1 will the minute and hour hand of a clock point in exactly opposite directions?

MODEL. The minute and hour hand point in exactly opposite directions at 6 o'clock. Within the next 12 hours they will point in exactly opposite directions 11 times, each time requiring $\frac{1}{11}$ of 12 h., or $\frac{1 \cdot 2}{1 \cdot 1}$ h., to get in this position. When they stand in opposite directions between 12 and 1, it is the sixth time they have so stood since 6 o'clock, and it is therefore 6 times $\frac{1 \cdot 2}{1 \cdot 1}$ h., or $\frac{7 \cdot 2}{1 \cdot 1}$ h., after 6; which makes the time $32\frac{8}{11}$ min. past 12. *Ans.* $32\frac{8}{11}$ min. past 12.

2. At what time between 7 and 8 will the hour and minute hand of a watch point in exactly opposite directions? At what time between 1 and 2?

3. A pasture is hired for $23. B puts in twice as many cows as C, and C 3 times as many as D. D's cows are in twice as long as C's, and B's 3 times as long as C's. How much should each pay?

4. G and H embarked in a speculation. G furnished $400 for 2 months, and received $120 for his share of the profit. H furnished $600 for 1 month, and received a proportionate share of the profit. What % of the whole capital was the profit?

5. A could dig a certain cellar in 12 days, B in 10 days, and C in 15 days. They all worked one day, when A quit. B and C worked the next day, when B quit. How long did it take C to finish it?

6. Ira's age is 6, Paul's 20; in how many years will Paul be twice as old as Ira?

MODEL. Paul is 14 years older than Ira; hence, when Ira is 14, Paul will be 28, or twice as old as Ira. But, as Ira is now 6, he will be 14 in 8 years. *Ans.* 8 yr.

7. In how many years will Stephen's age be half of Andrew's, if Stephen is 3 and Andrew 12?

8. Two horses trot in the same direction round a circular course 1 mile long. One goes at the rate of 6 miles an hour, the other 10. How many minutes after starting will they be together again?

9. Two brothers having different amounts of money, the elder equalized them by giving the younger as much as he already had. If the elder originally had $6000, how much had the younger?

NOTE. After the gift each brother had twice as much as the younger had at first; and, as the elder had given away as much as the younger had at first, he must have had at first 3 times as much as the younger.

10. Having 60 ducks in one coop and a smaller number in another, I put with the latter twice their own number from the other coop, and then found the numbers in the coops equal; how many ducks had I?

11. A man sold whiskey that cost $2.50 a gall. for $3. The price of this whiskey having advanced to $3, he watered it so that he could still sell it at $3 and make the same per cent. as before. How much water did he put with 10 gal. of whiskey?

12. A, having $\frac{1}{8}$ of a mile the start of B, in the next half mile he runs gains on B 10 rd. more; after which B runs 3 rd. to A's 2. How far in all did each run before A was overtaken?

13. There are 3 boxes, the first of which weighs 10 lb., which is $\frac{1}{4}$ of the weight of the second and third, while the third weighs $\frac{2}{3}$ as much as the other two. What is the weight of all three?

Quackenbos's Educational Works.

PRIMARY ARITHMETIC. Upon the Basis of the Works of George R. Perkins, LL. D. 1 vol., 18mo. Price, 22 cts.

ELEMENTARY ARITHMETIC. Upon the Basis of the Works of George R. Perkins, LL. D. 1 vol., 12mo. Price, 40 cents.

PRACTICAL ARITHMETIC. Upon the Basis of the Works of George R. Perkins, LL. D. 1 vol., 12mo. Price, 80 cts.
    Key to do. Price, 18 cents.

MENTAL ARITHMETIC. 1 vol., 18mo. Price, 35 cents.

HIGHER ARITHMETIC. 1 vol., 12mo. Price, $1.10.
    Key. Price, 65 cents.

FIRST LESSONS IN ENGLISH COMPOSITION. 12mo. Price, 80 cents.

ADVANCED COURSE OF COMPOSITION AND RHETORIC. 12mo. Price, $1.30.

PRIMARY HISTORY OF THE UNITED STATES; made easy and interesting for Beginners. Child's 4to. 200 pages. (Old edition.) Price, 80 cents.

ELEMENTARY HISTORY OF THE UNITED STATES. With numerous Illustrations and Maps. (New edition.) Price, 65 cents.

HISTORY OF THE UNITED STATES, for Schools. Illustrated. 1 vol., 12mo. Price, $1.30.

AMERICAN HISTORY, for Schools. Price, $1.25.

PRIMARY GRAMMAR OF THE ENGLISH LANGUAGE. 18mo. Price, 45 cents.

ENGLISH GRAMMAR. 12mo. Price, 80 cents.

NATURAL PHILOSOPHY, embracing the most Recent Discoveries. 12mo. Price, $1.50.

ILLUSTRATED LESSONS IN OUR LANGUAGE; or, How to Speak and Write Correctly. 1 vol., 12mo. Cloth. Price, 55 cents.

ILLUSTRATED SCHOOL HISTORY OF THE WORLD, from the Earliest Ages to the Present Time. Accompanied with numerous Maps and Engravings. By John D. Quackenbos, A. M., M. D. 1 vol., 12mo. Half bound. Price, $1.50.

D. APPLETON & CO., 549 & 551 Broadway, New York.

Dictionaries of Modern Languages.

FRENCH.

JEWETT'S Spiers's French Dictionary. 8vo. Half bound (formerly $3.50). Price, $3.00.
—— School edition. 12mo. Half bound (formerly $2.50). Price, $2.00.

MASSON'S Compendious French-English and English-French Dictionary. With Etymologies in the French part; Chronological and Historical Tables, and a List of the Principal Diverging Derivations. One 16mo vol. of 416 pages. New, clear type. Half bound. Price, $2.00.

MEADOWS'S French-English and English-French Dictionary. Revised and enlarged edition. 1 vol., 12mo. Price, $2.00.

SPIERS & SURENNE'S Complete French-and-English and English-and-French Dictionary. With Pronunciation, etc. One large 8vo volume of 1,490 pages. Half morocco (formerly $6.00). Price, $5.
—— Standard Pronouncing Dictionary of the French and English Languages. (School edition.) Containing 973 pages, 12mo. New and large type. Price, $2.50.

SURENNE'S French-and-English Dictionary. 16mo. 568 pages. Price, $1.25.

GERMAN.

ADLER'S German-and-English and English-and-German Dictionary. Compiled from the best authorities. Large 8vo. Half morocco (formerly $6.00). Price, $5.00.
—— Abridged German-and-English and English-and-German Dictionary. 840 pages, 12mo. Price, $2.50.

ITALIAN.

MEADOWS'S Italian-English Dictionary. 16mo. Half bound. Price, $2.00.
—— The same. A new revised edition. Half bound. Price, $2.50.

MILLHOUSE'S New English-and-Italian Pronouncing and Explanatory Dictionary. Second edition, revised and improved. Two thick vols., small 8vo. Half bound. Price, $6.00.

SPANISH.

MEADOWS'S Spanish-English and English-Spanish Dictionary. 18mo. Half roan. Price, $2.50.

VELASQUEZ'S Spanish Pronouncing Dictionary. Spanish-English and English-Spanish. Large 8vo vol., 1,300 pages. Neat type, fine paper, and strong binding in half morocco. Price, $6.00.
—— Abridged edition of the above. Neat 12mo vol. 888 pages. Half bound. Price, $2.50.

D. APPLETON & CO., 549 & 551 Broadway, New York.

Ollendorff's New Methods of Learning Languages.

FRENCH.

NEW METHOD OF LEARNING FRENCH. Edited by J. L. Jewett. 12mo. Price, $1.10.

METHOD OF LEARNING FRENCH. By V. Value. 12mo. Price, $1.10.
  Key to each volume. Price, 85 cents.

FIRST LESSONS IN FRENCH. By G. W. Greene. 18mo. Price, 65 cents.

COMPANION IN FRENCH GRAMMAR. By G. W. Greene. 12mo. Price, $1.10.

GERMAN.

NEW METHOD OF LEARNING GERMAN. Edited by G. J. Adler. 12mo. Price, $1.10.
  Key to do. Price, 85 cents.

NEW GRAMMAR FOR GERMANS TO LEARN THE ENGLISH LANGUAGE. By P. Gands. 12mo. Price, $1.30.
  Key to do. 12mo. Price, 85 cents.

ITALIAN.

NEW METHOD OF LEARNING ITALIAN. Edited by F. Foresti. 12mo. Price, $1.30.
  Key to do. Price, 85 cents.

PRIMARY LESSONS. 18mo. Price, 65 cents.

D. APPLETON & CO., 549 & 551 Broadway, New York.

Cornell's Educational Works.

FIRST STEPS IN GEOGRAPHY. 1 neat vol., prettily illustrated. Price, 40 cents.

PRIMARY GEOGRAPHY. Revised edition. Forming Part I. of a systematic series of School Geographies. Beautifully illustrated. Small 4to. Price, 65 cents.

INTERMEDIATE GEOGRAPHY, with New Maps. Revised edition. Price, $1.30.

THE SAME. Old edition. Price, $1.10.

GRAMMAR-SCHOOL GEOGRAPHY. 1 vol., 4to. Price, $1.50.

GRAMMAR-SCHOOL GEOGRAPHY. Revised edition. Large 4to, 120 pages. 31 pages of Maps. Price, $1.50.

PHYSICAL GEOGRAPHY. Price, $1.40.

HIGH-SCHOOL GEOGRAPHY AND ATLAS. Revised edition. Price, $2.60.

HIGH-SCHOOL GEOGRAPHY, separate, 1 vol., 12mo. Price, 85 cents.

HIGH-SCHOOL ATLAS, separate. Price, $1.70.

CARDS FOR THE STUDY AND PRACTICE OF MAP-DRAWING. 4to. In printed cover. Price, 45 cents.

SERIES OF OUTLINE MAPS, in large size, well printed on fine stout paper, and all mounted on fine muslin, neatly put up in a Portfolio, and accompanied with a complete Key for the Teacher's use. Price, $13.25.

SINGLE MAPS, each 90 cents.

DOUBLE MAPS, each $1.75.

THE KEY, separate. Price, 50 cents.

D. APPLETON & CO., 549 & 551 Broadway, New York.

Works of Richard Anthony Proctor.

OTHER WORLDS THAN OURS: The Plurality of Worlds, studied under the Light of Recent Scientific Researches. With Illustrations, some colored. 12mo. Cloth. Price, $2.50.

LIGHT SCIENCE FOR LEISURE HOURS. A Series of Familiar Essays on Scientific Subjects, Natural Phenomena, etc. 1 vol., 12mo. Cloth. Price, $1.75.

ESSAYS ON ASTRONOMY. A Series of Papers on Planets and Meteors, the Sun, Stars, etc. With 10 Plates and 24 Wood Engravings. 1 vol., 8vo. Cloth. Price, $4.50.

THE MOON: her Motions, Aspect, Scenery, and Physical Conditions. With Three Lunar Photographs, and many Plates, Charts, etc. 1 vol., 8vo. Cloth. Price, $5.00.

THE EXPANSE OF HEAVEN. A Series of Essays on the Wonders of the Firmament. 1 vol., 12mo. Cloth. Price, $2.00.

OUR PLACE AMONG INFINITIES. A Series of Essays contrasting our Little Abode in Space and Time with the Infinities around us. To which are added Essays on the Jewish Sabbath and Astrology. 1 vol., 12mo. Cloth. Price, $1.75.

D. APPLETON & CO., 549 & 551 Broadway, New York.

EIGHTEEN
CHRISTIAN CENTURIES.

By the Rev. JAMES WHITE,
Author of a "History of France."

1 vol., 12mo. . . . . Cloth, $2.00.

CONTENTS.

I. Century.—The Bad Emperors.—II. The Good Emperors.—III. Anarchy and Confusion.—Growth of the Christian Church.—IV. The Removal to Constantinople.—Establishment of Christianity.—Apostasy of Julian.—Settlement of the Goths.—V. End of the Roman Empire.—Formation of Modern States.—Growth of Ecclesiastical Authority.—VI. Belisarius and Narses in Italy.—Settlement of the Lombards.—Laws of Justinian.—Birth of Mohammed.—VII. Power of Rome supported by the Monks.—Conquests of the Mohammedans. —VIII. Temporal Power of the Popes.—The Empire of Charlemagne. —IX. Dismemberment of Charlemagne's Empire.—Danish Invasion of England.—Weakness of France.—Reign of Alfred.—X. Darkness and Despair.—XI. The Commencement of Improvement.—Gregory VII.—First Crusade.—XII. Elevation of Learning.—Power of the Church.—Thomas à Becket.—XIII. First Crusade against Heretics.— The Albigenses.—Magna Charta.—Edward I.—XIV. Abolition of the Order of Templars.—Rise of Modern Literature.—Schism of the Church.—XV. Decline of Feudalism.—Agincourt.—Joan of Arc.— The Printing-Press.—Discovery of America.—XVI. The Reformation. —The Jesuits.—Policy of Elizabeth.—XVII. English Rebellion and Revolution.—Despotism of Louis XIV.—XVIII. India.—America.— France.—Index.

From the Home Journal.

"In no single volume of English literature can so satisfying and clear an idea of the historical character of these eighteen centuries be obtained."

From the Providence Press.

"In this volume we have *the best epitome of Christian history extant*. This is high praise, but at the same time *just*. The author's peculiar success is in making the great points and facts of history stand out in sharp relief. His style may be said to be *stereoscopic*, and the effect is exceedingly impressive."

D. APPLETON & CO., 549 & 551 Broadway, New York.

The Life of Daniel Webster,

By GEORGE TICKNOR CURTIS.

Illustrated with Elegant Steel Portraits, and Fine Woodcuts of Different Views at Franklin and Marshfield.

In two vols., small 8vo. . . . . *Price,* $6.00.

OPINIONS OF THE PRESS.

From the London Saturday Review.

"We believe the present work to be a most valuable and important contribution to the history of American parties and politics."

From the New York Tribune.

"Of Mr. Curtis's labor we wish to record our opinion, in addition to what we have already said, that, in the writing of this book, he has made a most valuable contribution to the best class of our literature."

From the New York Journal of Commerce.

"A model biography of a most gifted man, wherein the intermingling of the statesman and lawyer with the husband, father, and friend, is painted so that we feel the reality of the picture."

From the Boston Post.

"Mr. Curtis has accomplished his labor with a fidelity that demonstrates how truly it was inspired by love. . . . This 'Life of Webster' is a monument to both subject and author, and one that will stand well the wear of time."

From the Boston Courier.

"It may be considered great praise, but we think that Mr. Curtis has written the life of Daniel Webster as it ought to be written."

From the Chicago Journal.

"I rejoice that the life and character of Webster are so large and so precious an inheritance to us all. Mr. Curtis has handled his task with judgment, and made an effective and exceedingly satisfactory book, one to take its unquestioned place with the invaluable memorials of American progress which we owe to Palfrey, Bancroft, and other American historical writers of the first rank."

From the St. Louis Republican.

"It is a work which will eventually find its way into every library, and almost every family."

D. APPLETON & CO., 549 & 551 Broadway, New York.

INTERNATIONAL SCIENTIFIC SERIES.

NOW READY.

I. FORMS OF WATER, in Clouds, Rain, Rivers, Ice, and Glaciers. By Prof. JOHN TYNDALL. 1 vol. Cloth. Price, $1.50.
II. PHYSICS AND POLITICS; or, Thoughts on the Application of the Principles of "Natural Selection" and "Inheritance" to Political Society. By WALTER BAGEHOT. 1 vol. Cloth. $1.50.
III. FOODS. By EDWARD SMITH, M. D., LL. B., F. R. S. 1 vol. Cloth. Price. $1.75.
IV. MIND AND BODY. The Theories of their Relations. By ALEXANDER BAIN, LL. D. 1 vol., 12mo. Cloth. Price, $1.50.
V. THE STUDY OF SOCIOLOGY. By HERBERT SPENCER. $1.50.
VI. THE NEW CHEMISTRY. By Prof. JOSIAH P. COOKE, Jr., of Harvard University. 1 vol., 12mo. Cloth. Price, $2.00.
VII. THE CONSERVATION OF ENERGY. By Prof. BALFOUR STEWART, LL. D., F. R. S. 1 vol., 12mo. Cloth. Price, $1.50.
VIII. ANIMAL LOCOMOTION; or, Walking, Swimming, and Flying, with a Dissertation on Aëronautics. By J. BELL PETTIGREW, M. D. 1 vol., 12mo. Illustrated. Price. $1.75.
IX. RESPONSIBILITY IN MENTAL DISEASE. By HENRY MAUDSLEY, M. D. 1 vol., 12mo. Cloth. Price, $1.50.
X. THE SCIENCE OF LAW. By Prof. SHELDON AMOS. 1 vol., 12mo. Cloth. Price, $1.75.
XI. ANIMAL MECHANISM. A Treatise on Terrestrial and Aërial Locomotion. By E. J. MAREY. With 117 Illustrations. $1.75.
XII. THE HISTORY OF THE CONFLICT BETWEEN RELIGION AND SCIENCE. By JOHN WM. DRAPER, M. D., LL. D., author of "The Intellectual Development of Europe." $1.75.
XIII. THE DOCTRINE OF DESCENT, AND DARWINISM. By Prof. OSCAR SCHMIDT, Strasburg University. Price, $1.50.
XIV. THE CHEMISTRY OF LIGHT AND PHOTOGRAPHY: In its Application to Art, Science, and Industry. By Dr. HERMANN VOGEL. 100 Illustrations. Price, $2.00.
XV. FUNGI; their Nature, Influence, and Uses. By M. C. COOKE, M. A., LL. D. Edited by Rev. M. J. BERKELEY, M. A., F. L. S. With 109 Illustrations. Price, $1.50.
XVI. THE LIFE AND GROWTH OF LANGUAGE. By Prof. W. D. WHITNEY, of Yale College. Price, $1.50.
XVII. MONEY AND THE MECHANISM OF EXCHANGE. By W. STANLEY JEVONS, M. A., F. R. S. Price, $1.75.
XVIII. THE NATURE OF LIGHT, with a General Account of Physical Optics. By Dr. EUGENE LOMMEL, Professor of Physics in the University of Erlangen. With 188 Illustrations and a Plate of Spectra in Chromo-lithography. Price, $2.00.
XIX. ANIMAL PARASITES AND MESSMATES. By Monsieur VAN BENEDEN, Professor of the University of Louvain. With 83 Illustrations. Price, $1.50.
XX. ON FERMENTATIONS. By P. SCHÜTZENBERGER, Director at the Chemical Laboratory at the Sorbonne. With 28 Illustrations. Price, $1.50.
XXI. THE FIVE SENSES OF MAN. By JULIUS BERNSTEIN, O. Ö. Professor of Physiology in the University of Halle. With 91 Illustrations. Price, $1.75.
XXII. THE THEORY OF SOUND IN ITS RELATION TO MUSIC. By Prof. PIETRO BLASERNA, of the Royal University of Rome. With numerous Woodcuts. 1 vol., 12mo. Cloth. Price, $1.50.
XXIII. STUDIES IN SPECTRUM ANALYSIS. By J. NORMAN LOCKYER. With Illustrations. 1 vol., 12mo.

D. APPLETON & CO., 549 & 551 Broadway, New York.

Novel Scholastic Method for the Colloquial Acquisition of Foreign Languages.

In course of publication, in 12mo.

THE MASTERY SERIES

FOR

LEARNING LANGUAGES ON NEW PRINCIPLES.

By THOMAS PRENDERGAST,

Author of "The Mastery of Languages; or, the Art of speaking Foreign Tongues idiomatically."

This method offers a solution of the problem, How to obtain facility in speaking foreign languages grammatically, without using the GRAMMAR in the first stage. It adopts and systematizes that process by which many couriers and explorers have become expert practical linguists.

Already published.

HAND-BOOK OF THE MASTERY SERIES; being an Introductory Treatise. Price, 50 cents.
THE MASTERY SERIES (German). Price, 50 cents.
THE MASTERY SERIES (French). Price, 50 cents.
THE MASTERY SERIES (Spanish). Price, 50 cents.

To be followed shortly by

THE MASTERY SERIES (Hebrew). Price, 50 cents.

"To gain a thorough command of the common phrases which the majority use exclusively and all men chiefly, is the goal at which the Mastery System aims, and we think that goal can be reached by its means more easily and in a shorter time than by any method yet made known."—*Norfolk News.*

"Mr. Prendergast's scheme has the merit of simplicity, being nothing more nor less than a deduction from the natural method pursued by children, aided by the reason or intelligence which children do not possess."—*Greenock Advertiser.*

". . . En un mot, c'est le système le plus pratique que la philologie ait produit pour l'enseignement des langues étrangères."—*L'Impartial de Boulogne.*

D. APPLETON & CO., 549 & 551 Broadway, New York.

Ollendorff's New Method of Learning Languages.

SPANISH.

A NEW METHOD OF LEARNING TO READ, WRITE, and Speak the Spanish Language, after the System of Ollendorff. By Mno. Velázquez and T. Simonné. 1 vol., 12mo. 560 pages. Price, $1.30.

KEY TO THE EXERCISES IN THE NEW METHOD OF Learning to Read, Write, and Speak the Spanish Language, after the System of Ollendorff. By M. Velázquez and T. Simonné. 1 vol., 12mo. 174 pages. Price, 85 cents.

NUEVO MÉTODO PARA APRENDER Á LEER, HABLAR, y Escribir el Español, segun el Sistema de Ollendorff. Para uso de los Alemanes. Arreglado por D. H. Wrage y H. M. Monsanto. (Neue Methode die Spanische Sprache lesen, sprechen und schreiben zu lernen, nach dem Ollendorff'schen System.) 1 vol., 12mo. Price, $1.50.

CLAVE DEL ANTERIOR. Price, $1.00.

MÉTODO PARA APRENDER Á LEER, ESCRIBIR Y HAblar el Inglés, segun el Sistema de Ollendorff. Por Ramon Palenzuela y Juan de la C. Carreño. Un tomo de 457 páginas, en 12°. Price, $1.50.

CLAVE DE LOS EJERCICIOS DEL MÉTODO PARA aprender á Leer, Escribir y Hablar el Inglés, segun el Sistema de Ollendorff. Por Ramon Palenzuela y Juan de la C. Carreño. Un tomo de 111 páginas, en 12°. Price, $1.00.

UN MÉTODO PARA APRENDER Á LEER, ESCRIBIR y Hablar el Frances segun el Sistema de Ollendorff. Por Teodoro Simonné. Un tomo de 341 páginas, en 12°. Price, $1.50.

CLAVE DE LOS EJERCICIOS DEL MÉTODO PARA aprender á Leer, Escribir y Hablar el Frances segun el Sistema de Ollendorff. Por Teodoro Simonné. Un tomo de 80 páginas, en 12°. Price, $1.00.

D. APPLETON & CO., 549 & 551 Broadway, New York.

PRIMERS
IN SCIENCE, HISTORY, AND LITERATURE.

18mo. . . Flexible cloth, 45 cents each.

I.—Edited by Professors HUXLEY, ROSCOE, and BALFOUR STEWART.

SCIENCE PRIMERS.

Chemistry........H. E. ROSCOE.	Astronomy......J. N. LOCKYER.
Physics.....BALFOUR STEWART.	Botany..........J. D. HOOKER.
Physical Geography, A. GEIKIE.	Logic............W. S. JEVONS.
Geology..............A. GEIKIE.	Inventional Geometry, W. G. SPENCER.
Physiology..........M. FOSTER.	Pianoforte...FRANKLIN TAYLOR.

Political Economy.........W. S. JEVONS.

II.—Edited by J. R. GREEN, M.A.,
Examiner in the School of Modern History at Oxford.

HISTORY PRIMERS.

Greece............C. A. FYFFE.	Old Greek Life, J. P. MAHAFFY.
Rome............M. CREIGHTON.	Roman Antiquities, A. S. WILKINS.
Europe.........E. A. FREEMAN.	Geography......GEORGE GROVE.

History of Europe........E. A. FREEMAN.

III.—Edited by J. R. GREEN, M. A.

LITERATURE PRIMERS.

English Grammar..R. MORRIS.	Shakespeare........E. DOWDEN.
English Literature, STOPFORD BROOKE.	Studies in Bryant....J. ALDEN.
	Greek Literature...R. C. JEBB.
Philology........ ......J. PEILE.	English Grammar Exercises, R. MORRIS.
Classical Geography, M. F. TOZER.	Homer........W. E. GLADSTONE.

(*Others in preparation.*)

The object of these primers is to convey information in such a manner as to make it both intelligible and interesting to very young pupils, and so to discipline their minds as to incline them to more systematic after-studies. They are not only an aid to the pupil, but to the teacher, lightening the task of each by an agreeable, easy, and natural method of instruction. In the Science Series some simple experiments have been devised, leading up to the chief truths of each science. By this means the pupil's interest is excited, and the memory is impressed so as to retain without difficulty the facts brought under observation. The woodcuts which illustrate these primers serve the same purpose, embellishing and explaining the text at the same time.

D. APPLETON & CO., Publishers, New York.

JUST PUBLISHED.

APPLETONS' SCHOOL READERS,

CONSISTING OF FIVE BOOKS.

By Wm. T. Harris, LL. D., Supt. of Schools, St. Louis, Mo.; Andrew J. Rickoff, A. M., Supt. of Instruction, Cleveland, O.; and Mark Bailey, A. M., Instructor in Elocution, Yale-College.

Appletons' First Reader............................	90 pages.
Appletons' Second Reader........................	142 "
Appletons' Third Reader..........................	214 "
Appletons' Fourth Reader........................	248 "
Appletons' Fifth Reader...........................	460 "

These Readers, while avoiding extremes and one-sided tendencies, combine into one harmonious whole the several results desirable to be attained in a series of school reading-books. These include good pictorial illustrations, a combination of the word and phonic methods, careful grading, drill on the peculiar combinations of letters that represent vowel-sounds, correct spelling, exercises well arranged for the pupil's preparation by himself (so that he shall learn the great lessons of self-help, self-dependence, the habit of application), exercises that develop a practical command of correct forms of expression, good literary taste, close critical power of thought, and ability to interpret the entire meaning of the language of others.

The high rank which the authors have attained in the educational field and their long and successful experience in practical school-work especially fit them for the preparation of text-books that embody all the best elements of modern educative ideas. In the schools of St. Louis and Cleveland, over which two of them have long presided, the subject of reading has received more than usual attention, and with results that have established for them a wide reputation for superior elocutionary discipline and accomplishments.

Of Prof. Bailey, Instructor of Elocution in Yale College, it is needless to speak, for he is known throughout the Union as being without a peer in his profession. *His methods make natural, not mechanical readers.*

D. APPLETON & CO., Publishers, 549 & 551 Broadway, New York.

www.ingramcontent.com/pod-product-compliance
Lightning Source LLC
Chambersburg PA
CBHW031445160426
43195CB00010BB/854